Mt. St. Helens eruption
Ted Bundy- Green River Killer

White Military anarchist's
Indian War's   TITANIC
Bloody Bozeman Trail   Perfect Storm
Sitting Bull-Crazy Horse   Swiss Air Crash
Custer's defeat   Egypt Air Crash
                              John Kennedy Jr.
Volcanic Michael   Sand Creek Massacre   World Trade Ctr.
Eruptions Kennedy   Columbine Co.   Pentagon Attack
San Francisco Quake Atomic Bomb
Watts Riots   Billy The Kid
Rodney King Riots   Lincoln County   CIVIL
Forest fires-plane Wars- Geronimo   WAR
crashes-Serial   Cochise- Wyatt
Killers,Robert F.   Earp-Tombstone John F.Kennedy-Tornado
Kennedy   Boothill   Tim Alamo   Alley
                    Pancho   McVeigh
                    Villa

                    Aztec Pyramid Sacrifices

18th Century
Pirates

Passage To Ascension:

A Leper at the Gates of Rome

By

Daniel A. Masias

(Muh Sigh Uh)

WHERE THE NEXT DISASTERS AND TERRORISM WILL STRIKE THE U.S.

ISBN: 1-4033-3531-1 (E-book)
ISBN: 1-4033-3532-X (Paperback)

This book is printed on acid free paper.

1stBooks – rev. 4/9/03

"Hermes, Son Of Zeus And Maia
And Messenger Of The-Gods"

Sirius: The brightest star in the east from July to September. It is a
binary star system, consisting of Sirius A and B. It is 9 light years from,
the Earth

Table Of Contents

THE STATE OF MANKIND

A conscious volcanic force under the earth is affecting every human being, and we are not aware of it. Millions of years ago animals and plants consumed oxygen, nitrogen and other gases and water. The water came from hexagon snowflakes and the gases were explosive. They all died and volcanic heat cooked their remains into oil. Today this oil is causing great problems for mankind. Death of the past has already caused great wars in the present. Oil fuels the machinery of war and civilization. There is a cause and effect from under the earth! Throughout history people with influence to change history for the better, have been deposed, or murdered. Joan of Arc, Ghandi, Dr. Martin Luther King, John F. Kennedy, Robert F. Kennedy, Princess Diana and others have all met an early demise. Thus the term "The good die young." These people went against a volcanic law that is unknown, but known in the Bible. In Genesis, God put a curse on mankind, because of the misbehavior of Adam and Eve. Horrible people throughout history such as despotic kings and queens have lived long lives to spread their destruction and death. The Czar, Mao, Tojo, Hitler, Stalin, Pol Pot, Castro, Sadam etc. have all lived long lives that spread great death and destruction on mankind. In modern times people such as John Lennon and Phil Hartman, who preached understanding have vanished, Powerful magnetic magma plumes are working their way to the surface of the earth, causing large earthquakes in the north east U.S. Other plumes are at work around the world. NOTE I published my research in the newspaper. It was 100% correct in predicting the physical location of chaos. On page 78-80-81 there are two research papers that I published in the Gazette newspaper in Colorado Springs, on January 7, 1998 and July 28,2000. The January 7th research said that in the future, all over the world including the United States, disasters would take place on the 6 volcanic triangles. Two and a half years later on July 28, 2000, I published the follow up "public test" results of the original research paper. The public test of my discovery was 100% correct! This discovery cannot be ignored by mankind. Steps must be taken to better understand what is happening on planet earth. Our ancient ancestors tell us that beings from the stars left instruments under the earth and then returned to the cosmos. These instruments are hidden and cannot be seen. New astronomical discoveries in the last ten years tell us that matter can exist and yet be invisible. We are told that the entire vast universe and maybe other universes were created from absolutely nothing. It suddenly burst out of nothing into an enormous explosion of geometric proportions and as it became bigger more matter came into being. Our universe was born in a roiling cauldron of fire and explosive violence. Mankind is a part of this inherited violent beginnings. Volcanic and snowflake hexagons are part of our inheritance. Without the snowflake we cannot have water and without the volcano we cannot have the proper balances of nature that sustain life. There is much more to this story. this book will explain human violence and it's affect in the world.

Diamonds, gold, silver etc. are products of volcanoes. Other valued geologic products such as petroleum and coal are harvested by man. Like the old miners who placed parakeets in mines to warn of deadly gas, an unknown volcanic entity has sent these highly prized minerals to the surface through the action of volcanoes, to act as geologic testing sensors, like the parakeet. Sensing trouble on the surface of earth, 6 electromagnetic volcanic triangles around the world are responding. This model is similar to the premise of the film "2001-A Space Odyssey." Gold, silver, diamonds etc. act as the same mechanism as the black volcanic monolith, that alerts superior beings that activity is taking place on the earth. This is done from the volcanic moon! The mining of these materials alerts the mysterious volcanic collective that activity is taking place on the surface of the earth, and measures must be taken to protect itself. This offensive activity takes the form of 6 global volcanic triangles, that starts sending out diseases, disasters, genocides and wars. These human plagues are global in nature. The earth is protecting itself, like a dog scratching at flees. Our ancient ancestors tell us that mysterious beings from the stars left instruments buried under the earth, that could not be found. Modern archaeologists do not want to hear this, but it is true. Not listening to this discovery is a mistake.

Introduction

The mysterious history of mankind is a subject that has no shortage of ideas. Because the ancient history of mankind travels back in time thousands of years, speculation, misinterpretation and deliberate rejection of certain historical figures and their written works; has clouded our view of man's ancient history. The history of mankind has many ideas and theories about it's origin depending on where one lives on the earth; and the color of ones skin. The only true and correct manner in determining the real history of mankind is to search for a common denominator that encompasses ancient peoples, the physical elements of the earth and the life forms of the planet. Scientific knowledge of the solar system and the universe must also be taken into account. It was not until the end of the twentieth century and after the dawn of the space age, that extremely critical cosmic knowledge became available through the use of deep space probes. As a consequence the true history of mankind is now beginning to reveal itself. The astronomical discoveries that are being made at the end of the twentieth century were spoken about in ancient written artifacts, thousands of years ago. The cosmic discoveries now being revealed, do in fact agree with the ancient written records of mankind, and there in lies the true history of the earth and mankind.

In the book that I have written, I have barely scratched the surface of this subject. However I believe that I, am making a positive contribution to the understanding of mankind's history. Most of the statements that I have made in this book can be verified. Other statements that I have made are based on geological features of the earth, as well as mankind's ancient temple sites; that exist all over the world. Still other claims are based on the earths weather patterns of lightning, hail, hurricanes, cyclones and the earths features such as volcanoes; lakes of fire and other natural geological features. The contents of this book is not an attempt to convince one single person, to what I have written. The reason that I wrote this book was that it represented a sort of personal understanding about myself, that I have never been able to figure out;until I came to grips with who I am. All my life I have been having dreams about water, and I could never understand this. As a native of Colorado, which is a semi-arid state;I have never been around the oceans or large bodies of water. Why I have been having such dreams about water, I cannot say. Maybe this is because water is the product of hexagon snow flakes and this has played a very big part of my research. When I was a little kid, I remember my uncle gave me some of his U.S. Marine insignia. I always remember playing with the dark metal showing the earth. and some sort of guns or swords crossed on the earth. This always stuck in my mind. Now I know what this small symbol was trying to tell me. Just like the Marine corp. song "From the halls of Montezuma to the shores of Tripoli." Both of these locations are associated with the volcanic triangle areas of the earth where great disasters have taken place in the past and continue to this day. None of this is a coincidence. For instance, the latest disasters taking place in Jerusalem and other parts of the West Bank, were marked with rain failing on the disaster areas of the Holy Land, rain comes from hexagonal snowflakes and plays a central role in life. Later in the book I will explain the importance of water and how it has a hidden purpose. Like wise I will also talk about why children in America are killing.

The book The Sirius Mystery by Robert Temple was important in my understanding of mankind's very complicated history. If any one wants to read an excellent book on the true account of mans history, and does not mind getting into a great deal of detail; this book is an amazing work of art. Robert Temple completely un-masks the secret Egyptian worship of the star system Sirius. His book is even more important because most people have absolutely no idea that the ancient Egyptians as well as other peoples of the earth, from the past and from the present; have been worshipping superior beings from the star system Sirius.

All of the various subjects that I have discussed in this book can be found at the local library, as long as it is large enough to provide a good range of books and research material. All of my investigative and research material came from genuine copies of ancient Egyptian sacred texts, college history books, geology, geography and college science books. Other research material was taken from various periodicals and magazines such as The National Geographics, Discover, Newsweek, Life magazine Parade, as well as others. Other reference books were also utilized in the research and writing of this book. The sober reality is that all of this information is available at any good public library, however the reader must not have the mentality of a fifteenth century flat earther;and believe that the universe revolves around the earth and mankind. If the mind is closed it cannot function and a person might as well be mindless, or a child. When man lets societies institutions think for him, and he does not question the world around him; his thinking processes become myopic as well as rigid. It is at this point that the truth becomes the victim, and our human understanding of the world in which we live becomes quite opaque.

In the book that I have written, which is about our ancient ancestors; I am merely repeating their words that mysterious beings are responsible for the earth and mankind. The ancient written artifacts that proclaim the return of these beings, does in fact originate with man's ancient ancestors. Therefore in the book I am just the messenger, who has researched the subject of the ancient Sirius religion. In the investigation of this subject, I have found that there is a great deal of truth to the claims of our ancient ancestors. The physical facts as well as the history of mankind, speak for themselves! From The Words Of The Late Professor W.B. Emery, From His Book "Archaic Egypt" At a period approximately 3400 years before Christ, a great change took place in Egypt, and the country passed rapidly from a state of advanced neolithic culture with a complex tribal character to two well organized monarchies, one comprising the Delta area and the other the Nile valley proper. At the same time the art of writing appears, monumental architecture and the arts and crafts developed to an astonishing degree, and all the evidence points to the existence of a well-organized and even luxurious civilization. All this was achieved within a comparatively short period of time, for there appears to be little or no background to these fundamental developments in writing and architecture.

Through the course of this research book we will learn that the earth's oceans consist of billions of trillions of liquid snow flakes that all have a magnetic electrical charge. Like minature solar systems these water molecules which have their original shape as a hexagon, are all orbiting around each other and interacting in a chemical and magnetic dance, which creates weird scientific realities. When a hurricane forms and impacts the ocean these billions of trillions become a churning cauldron of chemically and magnetically, chaotic hexagons. This results in the great disaster that strikes our coasts. Humans interact the same way. The body is 70% water and we are literally walking magnetic hexagons, who react either positive or negative with other people, ideas, events and the whole range of human experiences. This is why we look so hard for the right person in life. Like the hexagon bee hive that is a model of cooperation we humans are very happy when we find the right person whose magnetic hexagon aura matches ours.

If we are lucky enough to find our significant other, we know it right away when the feeling is mutual. The human magnetic hexagon can also be a negative quality. When people are at malls and in large crowds where every one is taking a separate direction, the human magnetic hexagon does not want to make eye contact and everyone sort of keeps their magnetic distance. Alcohol breaks this magnetic hexagon self imposed boundary, in a bar. It also causes the mob mentality to break free. In situations of great stress such as war, disaster etc. mobs of people gather together in a common cause as each individual magnetic hexagon person is excited to the point of frenzy, like bee's in a hive, and the angry waves in a hurricane the mob of people surges back and forth as they attack and counter attack. This is the same effect in war and many other human endeavours. This magnetic hexagon model has been carried to the extreme throughout man's history and has resulted in horrible tragedy. The most recent was the Nazi model. This ultimate human tragedy is the best example when human magnetic hexagon's band together in a perverted orgy and the geometric megalith grows larger and larger and stronger. The only force that can stop it is another entity of greater magnetic human energy. Another example of this destructive geometric magnetic energy is the love triangle. Connect 3 hexagons and the outer boundary will form a triangle. In this relationship, death or destruction is very common. Serial killers are part of this equation. They are alway's loner's and deal in stealth. Their magnetic hexagon energy is different from the regular person. Why is it that when we see the snow coming down, we want to go out and build a snow man. Snow flakes are hexagon in shape. The snow man is made of hexagon or water, just like humans. Once human hexagons are agitated, it is hard to stop the momentum that often follows, thus the mob mentality and common sense can cease to exist. The human road to destruction often follows. This is why we have laws and why the Pharaoh performed the annual ceremony of order. Chaos was feared by the Pharaoh, and if this ceremony was not performed, disaster could strike Egypt. The God's had created civilization and Egypt, from out of the great abyss of chaos. Chaos still rules the universe in a geometric fashion. This is the condition of mankind today, a controlled geometric chaos. At times this chaos breaks free of constraints and that is when great disaster, take place. This is the nature of war. What appears to be a great foggy chaos is actually very geometric and organized in it's nature. The result is that somebody wins and somebody looses.

HOW I DISCOVERED THE GREATEST GLOBAL ARCHAEOLOGICAL DISCOVERY IN THE HISTORY OF MANKIND

Mars has 4 giant volcanoes that form a triangle. Mars was the God of war to the ancients. I brought the Martian volcanic triangle to the Earth and connected all of Earth's major volcanic, earthquake and temple sites all over the globe. 6 gigantic triangles formed all over the Earth! On these triangles mankind's major wars, genocides and natural and manmade disasters have been taking place forever!

THE QUESTION OF THE MESSIAH AND MARS IN ROMAN TIMES

2,000 years ago around 6 B.C. when the Messiah was born in the Holy Land, Mars approached the Earth and was an amazing sight in the night sky. 2,000 years later during the summer of 2003 on September 1, Mars will again approach the Earth and shine at magnitude –2.9. It will outshine mighty Jupiter. This incredible sight has not been seen by humans for 2,000 years, since the birth of the Messiah. 2,000 years later my research has discovered why man is plagued by disasters! My real name is Daniel Masias (Muh Sigh Uh) and like the original Messiah who was born in Asia, my name also carries the reference to Asia M asia s. The Messiah of 6 B.C. lived under brutal Roman rule and I have lived under Roman rule as well. As a native of Colorado, I have lived under Love, Lamm, and Roman rule-Governors John A. Love, Dick Lamm, and Roy Romer. Maia is also in my name. She had a baby with Zeus whose name was Hermes, messenger of the God's! I also have the same birthday as the Pope May 18th. The Moon is the astronomical symbol of the Messiah whose dark volcanic ranges always face the Earth, telling man that there are 6 volcanic disaster triangles on the Earth. The Gods want man to look at them so he can understand what is happening on Earth.

ANOTHER MORE ASTOUNDING BOOK IS FOR SALE

A second even more amazing research book pamphlet is for sale at this web site. My second book is quite simple in its presentation, but the information that it contains is overwhelming in its scope and has major implications for all of mankind. Contact the above web site or write to THE GLOBAL ARCHAEOLOGICAL RESEARCH CENTER OF COLORADO P.O. BOX 14 GREEN MOUNTAIN FALLS COLORADO 80819.

ASTRONOMICAL SIGNS

Like the approach of Mars in 6 B.C. announcing the birth of the Messiah and the approach of Mars in 2003 heralding a global archaeological discovery, here in Colorado many meteors have been sighted in the night sky, and many earthquakes have taken place. In the spring of 2002 the planets lined up to form a triangle, announcing my global archaeological discovery. This is partly what my second book is about. There are two versions of this book pamphlet. One is plain and sells for $40.00 and the other is fancy and sells for $80.00. Each book has a certificate of authenticity and a Muh Sigh Uh signature. Proceeds will go to the GLOBAL ARCHAEOLOGICAL RESEARCH CENTER OF COLORADO. Checks can be made out to Daniel Masias. Each book pamphlet is personally hand printed and bound one at a time so allow 4 weeks delivery time.

The Greatest Global Archaeological Discovery In The History Of Mankind.

The book that I have written "Passage to Ascension" Is a research book about a passion that is very dear to me. I am an independent archaeologic-geologic researcher from Colorado. Since I was a little boy in La Junta and Pueblo Colorado, I can remember my parents Daniel and Helen buying and showing me books about the ancient Etruscans, Romans, Greeks and the Egyptians. Instead of reading comic books and the newspaper funnies, I spent hours reading and thinking about these ancient peoples. Even though we were a desperately poor family, my folks were able to spend nickles and dimes to buy these books at the local second hand book stores. My mother loved the Etruscans and my father was always talking about Aristotle, Plato, and Socrates. Both of them instilled in me a great love of these peoples. In the pursuit of truth about history, I learned to look at history in a different manner. Early on I learned to ask questions about our ancient ancestors and why they acted the way they did. My folks challenged me to answer my own questions. Human interaction and history fascinated me. As I spoke earlier, the human mental geometry of hexagons has everything to do with history and civilization. When civilization and people are working properly as it should, there is discipline and most geometric hexagon people work as one unit with a common belief and goal. This is very much like the bees in a hexagon hive. When chaos descends on humans or bees, both can be deadly. Everything in our world has to do with ordered geometry which equals peace. Disordered geometry equals chaos and disaster. The government, military, police business etc. hate disorder. They see discipline as the key to survival. Nature is the same way: When water vapor freezes onto the windows and surfaces of a car, beautiful flower patterns emerge. Disorder and harm comes to humans when salt is added to water. People cannot drink sea water. In Israel the Dead Sea and the volcanic triangle in the Golan Heights, causes humans great disaster, chaos and death. The hexagon influence of geometry is everywhere. I have been studying our ancient ancestors for many years, and the book that I have researched and written is a partial culmination of my studies and the dreams that I have had concerning them. I am thoroughly convinced that my research has uncovered an astounding global archaeological discovery. The physical facts concerning my research and the historical events that have occured prove beyond a doubt that my research and it's conclusions are correct. The information that I have uncovered has profound global implications for all of mankind. I see my research as a benefit to humanity. It is only through the sincere and honest global analysis and correct understanding of man's history and the geologic structures of the earth, that another correct picture of mankind emerges. There are profound mysteries surrounding humanity and the earth. Not much progress has been made in solving numerous perplexing mysteries, when it comes to archaeology. The earth harbors a wealth of mystifying ancient civilizations that have defied satisfactory explanation. Archaeologist's don't even know the names of some of these dead civilizations and what they called themselves.

It has become apparent that in mankinds ancient past, many secrets held by the rulers in power have affected the history of man. The Egyptians, Sumerians, Minoans Greeks, Dogon and, others all held very secret knowledge about beings from other world's. This knowledge was not conveyed to the average person on the street, but was kept as secret sacred knowledge by a few select high priests. The Pharaoh and his family were also aware of this knowledge. These ancient peoples all believed that outside the realms of the kingdoms, chaos and great destruction reigned. It was the job of the Pharaoh and others to make sure that the proper duties were performed every new year, to ensure that the new year would be prosperous and free of disaster. These rituals were performed to keep the forces of chaos at bay. This was very important to these ancient peoples. Some of this great knowledge was destroyed at the library of Alexandria, when it was burned down by the Romans and others. I believe history has shown that when rulers or governments withhold crucial and important information from the public, very often great tragedies, disasters and genocides have taken place. In our modern times, people around the world want to know the truth about UFO's and the public has a legitimate right to this information. If the truth is kept from the public, the chances of disaster for mankind only increases.

In modern times the last 200 years of global archaeology have been marked by a few great discoveries. One of these events was the deciphering of Egyptian hieroglyphs by Jean Champollion and the discovery of Troy by Heinrich Schliemann. Another great discovery was the tomb of Tutankhamen by Howard Carter and the discovery of the Dead Sea Scrolls by an Islamic Bedouin. Howard Carter's discovery of the Pharaohs tomb raised the specter of a curse. Word began to spread that the opening of the tomb had unleashed a curse on those who had disturbed the sleeping king. Since the beginning of the 18th century, British, French, German, Italian and American archaeologists were all excavating tombs in Egypt. From the 15th to the 18th century, Europeans were using ground up mummies as a medicinal potion! From the 15th to the 18th century Great disasters of every sort struck European society. The greatest discovery of all was Tutankhamen's tomb in February 1923 by Carter. What followed next was the greatest and most destructive world war in human history. World War II began 16 years after Howard Carter and others entered the tomb. This was not a fluke of some sort but a real cause and effect, because on April 15,1912 the Titanic carrying the mummy of Queen Taheb sank in the Atlantic ocean. These two events were related to the curse of the Pharaohs. After the sinking of the Titanic, World War I began in Europe. What followed was an astounding orgy of death and destruction. All of the World War I and II combatants, were the same archaeologists who were digging up the tombs in Egypt from 1900 to 1923. These people laughed at the idea of a curse. There is no doubt that the greatest excavations in Egypt in the early part of the century coincided with the greatest World War disasters in human history. This was not a coincidence but a true cause and effect. The idea of a curse on mankind is not a new idea. In the old Testament a curse is declared by God on mankind because of the misbehavior of Adam and Eve. This is the idea behind original sin. Every human being is born with original sin and of course Jesus died for our sins on the cross. This curse by God became a chilling reality when he banished Adam and Eve from Paradise and when Cain killed Abel. There is absolutely no question that all over the earth, brother has been killing brother for thousand of years, because of this curse.When we think about the Titanic and it's terrible fate we realize that there was much more to this story. The century began with the mistaken belief by mankind that it could challenge nature and by inference God himself. At that time there were Titans of industry such as Rockfeller and Carnegie who by their very nature believed they were invincible men whose word and deeds were God like.

Everything about these men and others like them were bigger than life. They believed that nothing was impossible to accomplish and they went out and built giant skyscrapers and great industrial mills and every sort of industrial complex, that made society what it was. In their mind nothing was not beyond their grasp. These people felt they were like Gods, and in fact they acted like the Titans of Greek mythology. The Titans were the second line of Gods, there were 12 of them, like the 12 tribes of Israel. These Titans of mythology were huge and that is why the Titanic was aptly named. This gigantic ship was the most luxurious modern and powerful ship of it's time and it was declared to be unsinkable by it's builders. The amazing ship was built in Belfast Ireland and the people who built her declared that she was the most beautiful ship ever built and nothing bad would ever happen to her. The very name of this great vessel the Titanic in fact was menat to conjure up the ancient Greek Titans and to place her in the company of the Gods. In the ancient mythology of these beings though, the Titans were not all powerful as they thought. One of the Titans sons Zeus led a successful rebellion against his father and became the leader of the pantheon of Gods. In a modern day lesson for humanity, the Titans were then imprisoned in a cavity in the underworld for the rest of eternity. Like the Titans who were deposed, the sinking of the great Titanic would set the stage for the 20th century. Soon World War I began and a great crisis of humanity took place. The Ku Klux Klan began to rise to prominence and in 1918 the Spanish flu suddenly came out of nowhere. This was the greatest pandemic since the 15th century of Europe. Mankind had challenged nature and the Gods and the beginning of the 20th century saw the greatest human disasters unfolding. The Pharaohs were right when they performed their annual rituals asking nature and the Gods to bless them for another year and to protect the people from the great dangers of chaos and disaster, which lurked right around the corner. There is no doubt that the great Titans of the 20th century, who believed that nothing was impossible, certaintly affected all of mankind.

With the advent of the great industrial age and all of man's modern advancements, it is no wonder that people felt the way they did. With the 20th century though there were ominous signs on the horizon, because the fast paced economy was on a fast track to a great economic crash that struck in 1929. When this great wall street debacle suddenly appeared, it had a very devastating effect on the common man in America and around the world. Suddenly the great Titans did not feel so powerful. There was now a sense that something was wrong. World War I had killed millions and a great pandemic had struck mankind, the Titanic had sunk, there was great labor unrest in America and huge marches on Washington. The Ku Klux Klan, was on the march and even nature and the Gods were against man. Astounding new ideas were being formulated. Alfred Wegner a meterologist declared that the crust of the earth was not solid but was actually floating over the earths surface on plates. At one time all of the continents were in one piece and that over millions of years they broke apart and floated off in different directions. Niels Bohr the Danish scientist and Albert Einstein were developing advanced theories in atomic theory, that would in a few short years be demonstrated to mankind. There were great changes going on in the world, but mankind was no longer feeling like the mythological Titans of Greek lore. More great disaster was on the horizon. A disaster so profound that to this very day, it sends chills down the spine. When humans sense a 6th sense of doom, there is always a cold chill that permeates the air. This feeling is always present when people have reported the presence of a malovelent ghost. The air gets cold and cold crystals form as in the film the Exorcist, where the room freezes over and there is the pure chill of evil. Soon the 20th century would experience an ice age of pure evil as the triangle or axis of evil began a campaign of world conquest.

VOLCANIC ACTIVITY IS AFFECTING

HUMAN BEHAVIOR

The earth's major volcanic, earthquake and temple sites around the planet are very interesting. Geologists tell us that volcanoes are responsible for creations of the earth. They are credited with giving rise to our oceans and our atmosphere. They also created the earths continents and other geologic formations. While these great enigma's have been beneficial to the creation of life on the planet, they have also had a split personality. Volcanoes have brought great destruction to the earth and it's people. Some of the greatest natural explosions have occured because of volcanoes. These gigantic and highly dangerous mountains are the source of some of mankinds most sought after commodities. Diamonds, gold and silver are expelled by volcanoes. In some cases the circumstances of their creation are different and not really understood. For thousands of years mankind has sought out gold and silver. These precious metals as they are known, have been sought out as sources of beauty, power and wealth. These materials have also been used for religious purposes, and as a result, untold peoples have labored and died because of these metals. Diamonds have not been immune to man's greed, and the untold human suffering they have caused. Everyone has heard of the curse of the Hope Diamond. Another famous diamond in this group is the Koh-i-Noor and the Regent as well as other stones who are famous for their cursed and bloody histories. Like gold and silver, diamonds have been the source of human greed, war death and destruction. It is very prophetic that volcanoes should be the source of materials that have caused such misery and death to the human species. There is a fourth source of material that mankind has a deep craving for. This commodity also comes from deep under the bowels of the earth. This material is cooked deep underground from the earth's intense heat and it is the flora and fauna of ancient earth, better known as oil. The modern world and it's economies cannot do without this material. Everything depends on the availability of this black liquid. In modern times war's have been fought over oil, and the entire world is involved in a tug of war over this material. Global geo-politics is now focused on how to keep the flow of this material from being uninterrupted. All over the world, oil companies are looking for this material, because the world's economies depend on it. At the present moment, the entire world is embroiled in a conflict, that is partially related to oil and religion, and terrorism. It is not hard to see that volcanoes and the interior of the boiling earth, have been sending up huge amounts of material that mankind considers to be very valuable and is willing to kill for it. We never think about this, but it is amazing how much influence volcanoes have over our lives and how much humanity has suffered because of them. People the world over have to drive their large SUV's and heat their big homes with oil, therefore humanity is now in great conflict, and the consequenses cannot be good for mankind. I am convinced that volcanoes and earthquake zones of the earth are responsible for unseen amazing influences, that have never been recognized before. These mysterious geologic forces that I speak of are highly intelligent, well hidden and are part of the evolutionary history of planet earth. I am not talking about some sort of made up creature from deep in the cavities of the earth. I am speaking of the earth as a living breathing entity, that knows what is going on at it's surface, and it has begun to react accordingly. The earth doesn't breathe or have vision like humans, but.it has an evolutionary consciousness that allows it to monitor itself, and mechanisms that give it the ability to react to certain conditions. The ancient Greeks believed that Gaia was the God that ruled mother earth. This is a rather ancient idea. Our ancient ancestors were correct as far as Gaia was concerned. However the facts show that there are some very mysterious and terrible forces that are affecting mankind. For all intents and purposes, these forces appear to be malovelent and sinister. History shows us this and every day the television and newspapers talk about the human disasters that these forces are responsible for. I am not talking about the obvious destruction that these forces unleash on mankind. I am speaking of an unseen and unrecognized global force that is highly developed and has precided over all life for the last 4 billion years. This global entity is so well hidden and pervasive, that it hides in plain sight. After thorough investigation and years of research, there is no doubt that this intelligence exists. In nature, many creatures have developed mechanisms that fool their prey. There are many species that have adapted various means of fooling and devising stratagies to protect themselves from predators. These defensive and offensive mechanisms are repeated throughout nature. My research tells me that these mysterious facets of nature, do not stop with the lower order of mammals, amphibians, fish and insects, bacteria and viruses. A global volcanic evolutionary mechanism is affecting the affairs of all mankind. Science knows that there exists throughout nature a system of checks and balances. This system works for the lower order of creatures on this planet, but what about the highest order of mammals on this planet. How has nature dealt with humans, who we believe to be the most intelligent beings on earth. The law is very simple and brutal. In order for life to continue something has to die. It is just that simple, there are no exceptions to this law. Mankind has always fooled himself into thinking that because he is the most intelligent being and can reason, therefore he is not subject to the same position as other

9

creatures. This has always been an arrogant attitude that some have had. Despite this, it has always been a well known fact that man can never win in a battle with nature. Nature has overwhelming awesome power over mankind and his constructions and his ideas. Over the centuries this power has been demonstrated. Even in modern times with the advent of advanced scientific equiptment, disasters have still taken place with great human suffering. Because of these sobering facts it is not hard to imagine that there are forces buried deep in the earth, that refuse to be exposed and keep a close pulse on it's domain. Mankind has always felt that nature was a monolithic force that followed a pre-ordained set of rules and laws. The sun rose in the morning and set in the evening. Winter was followed by spring, summer and fall.

BURIED INSTRUMENTS UNDER THE EARTH

Extremely ancient texts speak of mysterious instruments left under the earth for unknown reasons, that are unfindable. This text came from ancient Egypt and it's author is unknown to mankind. Is it possible that this could be true. If we believe this then we have to abandon the idea that mankind is the most intelligent being on the planet. Many people are not going to give up their old belief system. One idea that we don't have to give up is that nature holds raw brutal physical dominion over mankind. With the advent of modern instantaneous communications all over the globe, people can instantly follow and plot the locations of world-wide disasters. When we do this, if we have a little intelligence and insight, we can easily see that there are geometric patterns to human disasters all over the earth. When we go back into the history books and look for the physical locations of past war's, genocides and disasters, we see that the same fantastic geometric shapes are still there, all over the earth! Why is this happening to mankind and who started this? My Egyptian- Greek and geologic research has discovered an astounding global archaeological and geological discovery, but before I can delve into this matter, we must first consider some other relevent facts. These issues must first be discussed so that the reader is better able to understand what is read later on. The facts we must consider are wide ranging and profound. In the last 5,000 years of mankinds history, there have been 7 great civilizations that have developed around the globe. Four of these developed around the same time, and the others at varying times. The Tigris-Euphrates valley gave rise to Mesopotamian civilization. The Nile valley gave rise to the great Egyptian civilizatioin. The Indus valley gave rise to Indian civilization and China rose to power around 1500 B.C. These four great civilizations evolved into highly advanced socities that still influence mankind today. All of these civilizations have one common geologic characteristic. While some of them were near volcanoes, at the time that they rose to power, pyroclastic volcanoes were not erupting and dropping ash on their civilizations. Prevailing winds did not expose them to volcanic gases. As a result of this, these advanced socities while they were violent, did not venture out and conquer the world. The Greco-Roman, Japanese and Aztec civilizations were all different in their scope and directions. There were however some very common similarities. They were all physically located near highly active pyroclastic volcanoes. These civilizations were exposed to concentrated amounts of ash and volcanic gases. The ash fell in their fields where crops were grown. Volcanic gases saturated their civilizations and their drinking water was exposed to volcanic elements. These socities were highly advanced but were also very violent in their very natures. The Greeks were conquerors who fought among themselves. They were also capable of great constructions, philosophical thought and ideas as well as wonderful art

11

and music. Their Olympic games though were quite violent from the modern games of today. The ancient Romans were by far the most violent and expansionist empire of all. The modern world is abhored by the gladiatorial history of the Roman Republic. The Colosseum and it's ghastly history of death was located in a country that was highly volcanic. Capua was a few miles north of Vesuvius. This is where Roman gladiators and Spartacus were trained for the death games in the arena. The volcanic activity of Italy and the Roman empire went hand in hand. The entire geologic history of Italy and the Roman empire was and is today dominated by very explosive structures. During WW II Mt. Vesuvius erupted as it has erupted many times in the past. The hot springs of volcanic Italy were used by the Romans to maximum advantage. The ancient Romans were famous for their Roman baths and the aqueducts that brought water and snow melt to their great cities. The public baths, soaking pools and fountains of ancient Roman society are world famous. All of Roman society was built around their love of the baths and the great pleasures that it brought these people. How does one come to grips, that these people were highly advanced, yet they were very brutal and violent when they dealt with other people who were enemies of the state. The Roman Empire could not have existed without the forced use of human slavery. This entire society operated and thrived on the use of human slavery. Usually these slaves came from the defeated lands, that Roman soldiers occupied. There is no doubt that the highly volcanic lands of ancient Italy was very influential in the development of the Roman Empire. The Japanese home islands are famous for volcanic activity. Mt. Fuji is renowned for it's sheer beauty and the beautiful shimmering snow that covers this majestic volcanic mountain in Japan. This land is a tortured volcanic area that has many earthquakes and eruptions. This has made the Japanese people a very closed society. This civilization spawned the very violent secretive organization known as the Samurai Warriors. These fierce combatants are known as the people who used deadly long swords. The Japanese have been known as an expansionist empire. In their long history, they conquered large parts of Asia. During WWII they were able to defeat many of their enemies. It is an irony that these people have developed a very advanced civilization yet their history shows us that they have been extremely violent in their past history. Like the ancient Romans and their hot springs and baths, the Japanese are also famous for their baths and hot springs. These people love and revere the ancient traditions of their hot baths. It is understandable that in a highly volcanic land with erupting volcanoes, exposure to volcanic gases and ash, that the Japanese civilization would have been influenced by these mysterious volcanic forces. The land of the Aztecs and pyramids is full of highly explosive volcanoes. Archaeologic research has shown the Aztecs and other Meso-America natives to be very violent and agressive peoples. When Spanish Conquistadors entered the Aztec empire, it was like throwing gasoline on the fires of a volcano. Archaeological records show that the Aztec's were involved in ritual human sacrifice. Captured warriors from other tribes were taken to the top of a pyramid and with the use of a volcanic obsidian knife, the warrior's heart was cut out of his chest. The beating heart was held up for all to see. Eye witness accounts say that when white men saw this pyramid of death, the top of the structure was covered in human blood and reeked of human sacrifice. Is it any wonder that this highly volcanic land with erupting volcanoes, gases and ash, would spawn such a violent civilization? In a real sense the sacrificial pyramids of the Aztec's were awash in human blood, just as the Roman Colosseum was awash in the blood of human and animal sacrifices. There was really no difference in the outcome of the helpless victims. In the

Roman model, the sacrifice of victims was state sponsored for public entertainment and in the Aztec model, the human sacrifices were for the benefit of the God's. There is no doubt that highly volcanic regions of the earth have spawned advanced but very violent civilizations, but why?

AMERICA LAND OF THE 7 ACTIVE VOLCANOES AND THE BEAST

On May 18th 1980, on my birth date, Mt. St. Helens erupted with astounding force which resulted in many deaths. The eruption of this volcano in the Cascade range was not unexpected, but it's violence and far reaching effects were not expected. America has 11 volcanic peaks on the west coast. Only 7 of them are considered to be active and dangerous. What is unsual about these volcanic mountains is that they are all arranged in a curving bow from Washington state down to California. Mt. St. Helens however is visibly out front of the other 10 volcanoes in a westerly direction towards the Pacific ocean. My mother's name is Helen, this I believe is interesting, because this is what my' research and book is all about. Huge sections of the U.S. were exposed to ash and volcanic gases. Many children were exposed to these mysterious volcanic elements in the ensuing years. The eruption of this volcano deposited ash on the farming fields of America and the gases and other unknown chemical elements were injected into the fresh drinking water. This gigantic eruption was a watershed event in helping mankind to understand the unseen evolutionary processes that volcanoes exert on mankind and civilization. In the year's that followed this event, American society has noticed a marked increase in brutal and deadly violence by American teenager's. In past decades, violence was marked by the use of fists and brawls. Since the volcanic eruption of Mt. St. Helens, teenage violence is symbolized by the use of hand guns and machine guns at Columbine. Research articles, newspaper stories and magazines have described some American teenagers as devoid of human compassion for other people. The suffering's of other human beings do not bother these young people. They' lack empathy, understanding and a sense of brotherhood with other people. Modern entertainment for these young people has consisted of films, T.V. and video games showing great violence. Many people believe that our society has become more violent and immune to the sufferings of people in America and around the world.

Does this sound familiar. We have already discussed this earlier. The Greco-Roman world was very violent and they were not sympathetic to the sufferings of other peoples and animals. The Aztec's were the same, for they were involved in an orgy of human sacrifice and did not care about human suffering. Both of these great civilizations had no feelings for the plight of other peoples. Their purposes were for the subjugation of other peoples and their enslavement. There were no feelings of human compassion or empathy for the unfortunate victims. There was only the horror of being thrown into an arena while crazed blood thirsty people in the stands, yelled for your death. This was the reality for unfortunate people who lived in the lands of explosive volcanoes! For anyone who has ever asked the question of why these highly advanced peoples could be blood thirsty and yet enlightened in their philosophy and their democratic principles. Another question that we must ponder is more disturbing. Why did modern civilization copy the Greco-Roman model of government? There are a number of answers to this question. Some of them would be scholarly and based on the history of mankind, and other answers would be based on emotions, tradition and gut feelings. All would probably be correct to some degree. In any case, there is no changing the horrible history of mankind. We must look to change the plight of the human condition around the world, so that there is more equity for all oppressed peoples. In America the eruption of Mt. St. Helens has served to show man that there are forces under the earth that are affecting the affairs of humanity. This eruption served as a laboratory experiment that I always felt was a valid global conclusion. Any rational person who investigates the physical locations of the 7 great civilizations around the world, can easily see that explosive volcanoes do in fact exist among some of these socities. The question here is cause and effect. Is it a coincidence that the more violent socities were near these mountains or are there unseen geologic forces at work that affected their murderous attitudes? There is no coincidence here because if we look at the histories of these civilizations, they never deviated from their beliefs, so this is an indication that their attitudes were deeply ingrained. This is very indicative of a physical condition of the brain which we would call tradition. People have a belief system that is unshakable no matter how outlandish and destructive it is to them. It takes an open minded individual with education and common sense to change his mind and admit that their belief system may have been wrong. This person is able to change and has reached a more honest and fair conclusion. Since the eruption of St. Helens, great anguish has been expressed about the condition of American teenagers. There have been numerous school shootings and what appears to be senseless brutal violence and gang activity. People are afraid to venture out and they express fear of young people and gang activity. Some teenagers are accused of being remorseless for their crimes, as they sit in prison. Many people believe that our modern Western Civilization has become more violent.

This is a correct conclusion and it is related to unseen and unrecognized volcanic forces. In modern society people are more rude to one another and everyone wants to sue somebody. What are the prospects for increased violence in America due to volcanic activity? This is a question that many people would like to have an answer for. So far everything is ok. We must look to the ancient past for possible answers to this question. The Roman world was surrounded by active and dangerous pyroclastic volcanoes and their violent conquering civilization reflected this. In America there are 7 active and dangerous volcanoes and it has just been discovered that the megavolcano called the "Beast" in Yellowstone is an active volcano. This is an astounding and very troubling development for America and the world. Mankind has never witnessed the eruption of a mega-volcano with a crater 30 by 70 miles in diameter. Geologists are saying that this could endanger the very existence of all mankind. It's eruption would be comparable to being struck by an asteroid. Right now this beast is bulging the thin crust of the earth in the direction of Los Angles and San Diego. If it erupts this will be Armageddon as the Bible declares. So far nothing has happened and we must be concerned if numerous and moderate earthquakes begin to strike the area. The past volcanic history of America has been very violent. The ash that has been deposited has made America the bread basket of the world. Great amounts of rich volcanic ash has fallen on the great growing fields of America and this has created a great bounty for Americans and the world. There have not been enough active erupting volcanoes in America to cause her government to go the way of the Roman Republic, and it's murderous orgy of death. For instance, before WW I President Wilson and many Americans did not want to enter the war and they felt that it was an European war and none of our concern. At that time America had great power in her industrial might and her military. A powerful American President could have become a modern Caesar and set out to conquer the world. After all other humans around the world wanted to do this. But because there were no active volcanoes on the west coast and none had erupted in 6600 years since Mt. Mazama in California erupted, Americans were not influenced by volcanic forces, to go on the rampage and conquer the world. Wilson had refused to enter the European war. Later however he was forced to do so. America it seems is a good example of my global archaeological discovery that unseen volcanic forces are hurting humans. Since WW II when the U.S. had the atomic bomb, it could have easily conquered the rest of the world, but it didn't, because it was not under the influence of volcanic forces. In today's modern world many people around the world are upset that America has appointed itself the world's police man and they believe that she has too much political influence around the world. Many people around the world believe that America has in effect conquered the entire earth with her economic might and that she imposes her will on other people who do not agree with her. Because of religious conflict and differing beliefs, other's cite the emergence of terrorism in America. Because of the position that the U.S. has assumed after WW II and the end of the cold war, many people believe that the ghastly terrorism that struck the innocent people in New York city and elsewhere, was a pre-ordained event. There is no doubt that the unknown volcanic forces of the earth are today affecting what man does to man. The cause and effect is there in plain sight. For instance in modern astronomy today, many planets have been discovered outside of our solar system as well as black holes in the universe. These structures cannot be seen by astronomers but instead are infered to exist because of the way that they physically affect surrounding stars and interstellar matter. Nearby matter is gobbled up or turns dark and cannot be seen and unseen planets physically exert gravity on it's host star, causing the star to wobble in it's orbit. These examples are modern day scientific avenues of cause and effect. This type of scientific inquiry is used all the time. My research is no different from the methodology of astronomers in their discovery of planets and black holes. They are observing violent physical events in the universe and they cannot physically see what is causing this violent behavior therefore to correctly describe what they are viewing they have postulated that planets and black holes are to blame. The same can be said of radio waves, radiation, odors and heat waves to name just a few examples. All of these physical scientific principles are invisible to the naked eye. Radiation can be detected by a geiger counter, radio waves can be heard on a radio, odors can be detected by the nose and heat can easily be felt by the human body. Earlier I mentioned Alfred Wegener and his discovery of our floating or drifting continents. He observed that violentt physical events had taken place on the earth and he deduced that the continents are drifting. He was proven correct. My research is no different from all of the other scientists and researchers. I have researched and documented the affects of highly explosive volcanoes around the world and their influences on mankind and civilization. Even though volcanic gas and ash cannot be seen physically touching man, there is an useeen and unknown effect on the minds of men, that have been affecting civilization for thousands of years. This is the same kind of premise that astronomer's use in their research.

THE COMPOSITION OF THE EARTH AFFECTS MANKIND AND HIS CIVILIZATION

Why is the earth composed of certain materials and gases? Is it an accident or chance event that our planet is covered in water? Why do humans and other creatures breathe oxygen and nitrogen? Oxygen is an explosive gas and nitrogen is chemically related to nitroglycerin that is used in the manufacture of dynamite. Life on planet earth depends on a number of deadly dangerous and explosive gases and elements! The earth and all life except for a few ocean creatures depend on the deadly white radioactive light of the sun, which is 93 million miles away from earth. If the earth for unknown reasons were to get closer or further away, the earth would either burn up or freeze into a wasteland. Why is it that life hangs in the balance, and that any small fluctuation can bring almost instant total destruction! It seems as though the God's have placed mankind and civilization on some sort of evolutionary probation, to see how well he can handle adversity. One might even argue that mankind through Einstein and other's have been given the very essence of the God's. The very material that they are composed of, if you believe in mythology. But who say's mythology is myth. We have been taught to believe certain truths, but are they really valid truths? For instance, science at one time taught us that there was no water on Mars, New scientific investigations have proven the old science to be incorrect. This planet at one time had alot of flowing water on it's surface but an ancient cosmic disaster boiled off the surface seas. The point of my statement is that what we are taught to believe may not be the truth at all and we must not place ourselves in the mindset of rejecting all new ideas and concepts. If we do this we place our minds in a myopic condition and we develop tunnel vision, that will eventually imprision us. At the end of our day's when we are ready to take our last breath, if we are lucky enough to be without physical pain, and are able to think a coherent sensible thought, we will say "What was this all about?" The point is that nothing in this violent world that we live in is a given and the same is true of our solar system. Past experience teaches us that disaster is just around the corner. In the last few years there have been numerous reports of giant asteroids that have approached the earth and barely missed our planet.

In most cases these space mountains were seen as they passed the earth. No body knew they were heading for the earth. Scientists consider these misses a near bulls eye. One of the most recent near misses occured on January 7, 2002. A 1000foot asteroid flew past the earth at 68,000 miles per hour. Astronomers never saw the approach of the monster. It was only seen after it sped past the earth and could not be seen because it came from the direction of the sun. In the last 10 years there have been a number of unsettling close encounters with these giant boulders. In the last 5 years the appearance of 4 large asteroids that have come close to the earth, has frightened many people. The earth is approaching an area of interstellar space that is inhabited with more of these space mountains. The earth is in more danger of being hit." In 1908 Tunguska Siberia was struck by a large unknown space object. The resulting explosion was enormous and it's presence was felt around the world. To this day researchers still do not agree what this object was. Some believe it was a stony meteor and others say it was a comet. Still some scientists are not positive what it was. This object most likley was a natural plutonium meteor that detonated above the earth, when the high pressures of rentry caused it to detonate. In 1972 a tourist in Jackson Hole Wyoming photographed a 1,000 ton meteor as it streaked through the atmosphere. There is no doubt that on numerous occasions the earth and mankind has come extremely close to disaster. How long can this continue before a huge event takes place and it is too late. The common theme of mankind has been that life hangs in the balance. There are laws and rules that govern these kinds of issues, but mankind has not been able to discipher them. Just as there are many dangers lurking in the sky above the earth, these dangers are a microcosm from where the earth came from. By the same token, all life on the earth breathes explosive oxygen and nitrogen gases. This is a very bizarre type of existence. We literally breathe nitrogen gas which is related to dynamite and the weapons of war. Nobody but chemists ever thinks of this bizarre fact. Why do we breathe explosive gases and chemical elements related to war, death and destruction? We can remember when Apollo 11 burned. The oxygen atmosphere in the craft ignited because of an electrical short. The interior burned like a blow torch killing all 3 men. Oxygen is a very dangerous gas and open flames should never be present. Today people everywhere can be seen carrying their oxygen bottles, because they need the extra gas to survive. Oxygen can be a very scary gas. In July 1945 some nuclear scientists working on the Manhattan project at Alamogordo New Mexico were genuinely afraid and concerned that the first atomic bomb test would ignite the earth's rich oxygen-nitrogen atmosphere. This would have caused an incredible disaster in New Mexico. Earth's living plant's, trees, shrub's, etc. take in carbon dioxide and exhale oxygen. The earth's oceans are also repositories of carbon dioxide. The life and death cycle of the earth is a remarkable closed system. The various life systems of the planet work in unison and cooperate in such a manner that life continues. But why should plants give off explosive oxygen gas, when their survival is not really dependent on the activities of mankind. Man has only been on the planet for a short time and plants have been here for a much longer period of time, therefore man must have adapted to the atmosphere that he inherited. So we must conclude that the dinosaurs must have breathed oxygen and nitrogen and were subject to the same mysterious volcanic elements. Humans and animals have bodies that are literally saturated with oxygen and nitrogen as well as water. Human waste causes pollution to the land. Just like the alien monsters of film who drools toxic slime that eats through steel, human waste when concentrated does the same kind of damage. Vast amounts of nitrogen waste are released into the environment each year and this nitrogen pollution has spoiled many parts of the earth. Septic systems everywhere are responsible for ground water pollution because of the nitrogen. Is it any wonder that when humans breathe explosive gases, they all expel biologically toxic and explosive waste. This is also true of other creatures. Modern plumbing around the world is based on this scientific fact. Every modern plumbing system has a series of vent stacks in the roof of the home, in order to vent the explosive sewer gases that build up in the plumbing system. Modern cities have street sewer systems that vent sewer gases and drain off rain water. Most creatures on this planet except deep ocean tube worms who live near volcanic vents and live on the sulfur deposits, breathe a bizarre combination of explosive gases in order to survive.

EXPLOSIVE GASES KEEP ALL HUMANS ALIVE!

Is it any wonder that humans all over the earth are violent and cruel and engage in warfare. It only makes sense that if mankind and all other creatures live on explosive gases and their bodies are saturated with these chemical compounds, then there are going to be big problems in our world. Some one some where long ago decided that this planet was going to be a very bizarre place to live and because of somebodys decisions, that's the way it is. The very history of mankind proves that volcanic and biological forces that rule on this planet have been the driving forces behind history and every aspect of the human condition.

This is the reason that there have been countless war's, genocides and natural and man-made disaster's on this planet, and the reason that they are continuing right up to this very moment. For instance look at what is happening in Israel the West Bank and Egypt, at this very time.

In the land of the Prince of Peace a great disaster is now taking place. This holy land the land of Abraham, Moses; and Jesus, is bearing witness to a terrible human Palestinian tragedy. This is the land of very ancient volcanic structures and activity. The Golan Heights is a 3,000 foot tall and 40 mile long volcanic plateau that has witnessed a great deal of tragedy for 5,000 years. Again we have the mysterious volcanic forces at work. This is the land where the Messiah preached the word of God around the sea of Galilee. This is very prophetic for this body of fresh water is actually a small mega-volcano that has been filled in with water. Ancient texts and the Bible speak of fire and brimstone falling from the sky and this is exactly what volcanoes blast into the sky. Watch the film "The Ten Commandments" with Charlton Heston and the sacred mountain where God resides show's a rumbling brooding volcano with fire showing through the gray ash smoke. Even the Holy lands are not imune from the mysterious volcanic forces of the earth. There are also other elements of the earth that have important effects on mankind. The earth is 70% water and 30% land mass. The water vapor clouds of the eatth cover 60%, of the land mass at anyone time. Mankind is a carbon based species and are composed of 70% water and 30% bone structure. This is identical to the composition of the earth. Like the question of why plants produce oxygen for humans, when there is no biological reason to do so, except to keep humans alive so that they can make war, we must ask ourselves another question. The question is, why is water so important to all life on the planet, and why does fresh water always come from hexagonal shaped snow flakes and crystals? Snow flakes and their hexagonal shapes have been studied since the 16th century by scientists who felt this was a very important subject. Johannes Kepler in 1611 wrote a treastie titled "A New Year Gift or on the six cornered snowflake." This was the first scientific investigation to explain why

24

crystals always display a hexagonal sixfold symmetry. In 1635 Renee Descartes the French scientis-(I think, therefore I am, and God can be found in geometric shapes.) like Kepler discovered that snowflakes could take many shapes, but all were a variation on a few basic hexagonal shapes. Of all the mysteries surrounding mankind and ancient civilizations, why does fresh water, the liquid life blood of all life on earth, originate from frozen hexagonal snowflakes? On this planet it never fails that what can go wrong will go wrong. For instance many of mankinds inventions have been used for the waging of war. Even the very structures that we live in have been refined for the purpose of war such as the deeply buried bunker. Almost every scientific discovery and commercial application can and has been turned into some sort of military application. When mankind had a great chance to learn very secret and ancient knowledge about his past the Romans, Christians and savages burned down the library at Alexanderia. For hundreds of years Europe languished in perpetual religious dark ages while the Inquisition went on. Other parts of the world like China and the Middle East were more fortunate, but none of these socities could or would share their advancements. Bad luck has been dogging mankind for centuries. In modern times almost every scientific advancement has been turned into some sort of military application. To make matters worse, atomic energy was used to make nuclear bombs, which are now threatening all of humanity. Humanity has not been able to create a positive instrument or device that could greatly improve the condition of mankind. The only types of scientific watershed breakthroughs have been devices of war. This opens up an opportunity for philosophers and thinkers as well as researchers to give mankind a better direction. The condition that we find ourselves in is aptly and soberly demonstrated by the deadly Van Allen radiation belts that surround our planet and literally have mankind imprisoned on this planet. This takes us back to the question of what exactly does the hexagonal shape of the snowflake mean to humanity? Snowflakes and crystals form from water condensed from it's gaseous state and grow around a small nucleus, usually a speck of dust. As more and more water molecules attach to the crystal, they do so according to unknown laws and form the distinct hexagonal shape. If we look up the word hexagon in the dictionary we are told that this is a six sided geometric structure. By the same token the root word of hexagon is hex. This means an evil spell, curse or bad luck. Most people don't know that there are other examples of hexagons in nature. Only geologists know that there exists a hexagonal volcanic basalt. Yes that's right, this is associated with volcanoes and now we are returning to these mysterious mountains again. One of the large examples of this material in the world is located at Devil's Tower in Wyoming, hence the name Devil's Tower because of the hexagonal shape of the volcanic basalt. This material exists in huge columns that create the giant tower. At one time this tower was the interior of a volcano whose sides over time were eroded away. In California at Devil's post pile the hexagon" basalt material exists in much smaller sections the size of fence posts, thus the name Devil's post pile. There are only two other places in the world where large amounts of column sized hexagonal basalt exist. One location is Giant's Causeway in Northern Ireland and the othe is Nan Matol in Micronesia. When white men asked native Indians of Wyoming for the translation about the meaning of the Indian name for this tower, the white men thought they had a mistranslation. It was translated as "Bad God's mountain The explorers thought that they had gotten the translation wrong. Because of the hexagonal shape of the basalt the rest is history. At Nan Matol in Micronesia there exists a large abandoned temple site, that was constructed out of large hexagonal columns. Archaeologists don't know who constructed these mysterious structures. Nearby volcanoes and other geologic features provided this hexagonal material. In this part of the world all of the islands were formed by underwater volcanoes and there are still many active volcanoes here. The famous WW II photograph of Marines hoisting the American flag over Mt. Surabashi, a 600 ft. extinct volcano, gives us an idea of the kind of structures in this part of the world. Native peoples of this region have oral histories about this mysterious place that go back centuries. They do not know who built the temples and the native do not dare visit Nan Matol because they believe that it is haunted by evil spirits. At night strange lights and evil voices are heard. If we investigate this part of the world we realize that the warm ocean weather pattern called El Nino starts in the region of Nan Matol! When El Nino shows up in the waters of Peru, it is Christmas time, thus the name for the Christ child. El Nino sets in motion a series of global weather changes that causes great destruction around the world This weather change originates with the volcanoes of Nan Matol! The Christ child is also associated with snowflakes in Western civilization. Churches around the world depict hexagons on their ceilings and elsewhere. Many religious scenes show a hexagonal snow flake with the Christ child. The third location of large hexagonal volcanic basalt is located at Giant's Causeway in Northern Ireland. This region of the world has been the scene of thousands of years of wars and disasters on a large scale. From 800 years of English and Irish warfare to the out break of WW I and WW II. As I mentioned earlier, the Titanic was built in Belfast Ireland. Volcanic hexagonal material has existed in this part of the world for millions of years. There is no doubt that mysterious volcanic forces have been at work in Europe and have caused great human disaster and suffering over thousands of years. At Devil's Tower in Wyoming the hexagonal basalt has not been a stranger to human disaster. In the 18th century from Colorado to Washington state, great conflicts and Indian wars took place in this part of the American west. Sitting Bull, Crazy Horse" Chief Joseph and other Indian leaders, all fought to keep their ancestral lands. White prospectors had found gold on the Indian lands, and native Indians fought the U.S Army and the prospectors to keep their sacred lands. Scholars consider the Sand Creek massacre near La Junta Colorado to be one of the worst to occur. Other battles that took place in this area of the world was Wounded Knee, Chiet Joseph's Trail of Tears enroute to Washington and the Battle of the Little Big Horn. This was Custer's last stand with the 7th calvary. This volcanic region of the U.S. also witnessed the death of many pioneers on the Bloody Bozeman Trail in Montana. Before this area was settled, a great. amount of lawlessness existed in these volcanic regions of the west. Butch Cassidy and the Sundance Kid operated here. Wild Bill Hickock was killed in a saloon holding the famous dead man's hand and Annie Oakley showed off her daring shooting skills. All over this region there were many desperados involved in crime and serial killings, as well as cattle wars and feuds. In Utah the Mormon's were involved in armed conflict to save their religious freedoms. They raised an army to fight the U.S. government who wanted to subdue them. People died in armed conflicts in this part of the country, with volcanic structures nearby. In modern times, powerful nuclear tipped missles sit in underground silos in the South Dakota plains near Devil's Tower. These are the ultimate weapons of war, and are meant for the Chineese or Russian mainland. For a number of years now the Hell's Angels have been meeting in Sturgis South Dakota near Devil's Tower. In Montana and Idaho, a very volcanic region, white supremacists militia groups have openly defied the U.S. government and this resulted in Ruby Ridge. In 1988 a catastrophic forest fire struck Yellowstone where the mega-volcano Beast is located. In this part of America, the earth's crust is thin and there exists huge roiling reservoirs of liquid magma from Colorado to Washington state to Baja California and back to Colorado. This forms a triangle structure were all along it's boundaries great disaster have taken place for hundred of years. When the "Beast" last erupted in Yellowstone around 600,000 years ago, it killed standing heards of camels in Nebraska and did astounding damage to the U.S. and affected the entire earth. Now this mega-volcano is rising up like a baking cake and threatening to erupt. As terrible as volcanic eruptions can be for all life on the planet, they have played a central role in the development of life and civilization on this planet. This is also true of the water's of the earth. As I discussed earlier fresh water has been one of the necessities of life and this element has it's origins in the hexagonal snowflake. In the natural world that we live in, everything that exists has a geometric shape to it. Mankind uses geometric shapes in all of his constructions and his cities and farm. Get on a plane and look down at the earth and see all the geometry. The natural shapes of coastlines, mountains, deserts, plains, etc all around the earth also have natural curved geometric shapes, This is generally unrecognized by the lay man. Around the earth there exists deadly magnetic and radioactive belts that are doughnut shaped and have a very uniform geometric shape. Everything in our world is in some form of geometric shape, therefore we have the snowflake in a hexagon shape. Ancient texts tell us for those who are willing to listen that for some unknown reason our earth is called the unclean earth. Just what does this mean, is unclear in this ancient manuscript. During WWII in the south Pacific near Nan Matol, the location of hexagonal volcanic basalt and, mysterious temples, the greatest human battles took place on volcanic islands between the Japanese Army and the U.S. Navy and Marines. Throughout this whole region of the Pacific and it's deep waters, the greatest most deadly and tragic war took place. Like warfare that has been taking place all over the earth for thousands of years, this one was different because the island battles were all surrounded by water. This is the reference for the unclean earth, or the soiled waters of the earth. Mankind's blood has soiled the blue Pacific water's through his war's. Life on this planet has not been Utopia. The history of civilization on this planet has been one of uninterrupted war, genocide and natural and man-made disasters for thousand" of years. Some civilizations have been more destructive than others, depending on their proximity to explosive volcanoes The written histories of mankind such as the Bible, Indian texts, Plato's writing and other sources all speak of great disasters from the destruction of Atlantis to Sodom and Gomorah. In the case of Plato's Atlantis, most likely this great civilization was destroyed by an undiscovered mega-volcano. In Plato's writings we have the same theme of the Bible. Mankind misbehaves and then God decides to punish him with a worldwide curse. In Atlantis the people became greedy, arrogant and self absorbed so God punished them by sinking them into the waters. Can modern mankind avoid this ancient hexagonal curse? Violence is a very common theme in man's history. The Big Bang is the very beginning of our Universe. What ever you believe in as to how we came in to being through God or a natural state of evolution, There is no getting around the fact. that God or evolution started the universe with an awesome mega explosion, No matter who did this, why did God or whoever do it with such violence? Now we can understand why there is so much violence in the universe including the geometric violence on earth. As the universe blew up, it blew up in a geometric pattern, just like a regular explosion. Conclusion; our earth is a microcosm of the universe and that is why the 7 triangles of the earth exhibit such prolific violence and disaster. This is why insurance companies call disasters act's of God. This reminds me of the painting by William Blake. His painting depicts God leaning over the earth with his hand on a pyramid or triangle. This is what my research is about. Go outside at night near the end of the month and look at the Moon. You will see a very shiny volcanic body that is illuminating our planet. It exerts great influence on the earth and people and is the symbol of the Messiah. People have a term called lunatic for the violence associated with a full moon. The volcanic moon is telling us that God's curse in the form of triangles is affecting the entire planet. This message is in plain hidden sight up in the night sky. The very fact that life is sustained from water, and has it's origins in hexagonal snow crystals, tells us a great deal. We remember that school science books spoke of the emergence of past ice ages. We have read accounts, of the disaster that the ice ages have inflicted on life. This reminds me of the film "The Exorcist," in which the evil entity causes the entire room to freeze up. This freezing effect is not unlike the freezing of a hexagonal crystal flake. On another level the hexagonal animal magnetism of each human being is one of the determining factors in how we accept others.

WHO IS DANIEL MASIAS pronounced MUH SIGH UH real name

I was born on may 18, 1949. Pope John Paul was born on May 18th and Mt. St. Helens erupted on May 18th, which is the name of my mother Helen and my research is about volcanoes. Masias Maia is in my last name. Maia in Greek mythology had a baby with Zeus, King of the God's. This baby is named Hermes or Mercury, who is the Messenger of the God's. Asia is also in my last name. The Messiah was from the near East in Asia or the Levant. In Greek mythology Helen, my mother's name was a farorite of the God's. I am part Jewish and a very distant relative was a member of the Sanhedron Jewish-Hebrew court. The Chinese, Asia, consider May 18th to be a very good and sacred day. Like the Messiah who was a carpenter, I have built a number of homes and cabins with my bare hands, with no one helping me. I live in the home that I built by myself. My mother's father John was a sheep herder in Colorado. In the 20 volume Oxford English Dictionary, on the origin of words, under the Messiah, the names Daniel and Messias; are used to explain the Messiah. My dad named me Daniel Masias, this is the phonetic sound of Daniel Messias. The name Masias is pronounce Muh Sigh Uh. The Jewish people of Israel believe that the Messiah was born when Israel became a country on May 14, 1948. On May 15, 1948 the Arab countries around Israel attacked her. After much fighting, Israel did survive as a country. I was born on May 18th 1949 the exact time that the Messiah was supposed to be born. The astronomical sign of the Messiah is the Moon, and the dark splotches of the Moon are volcanic plains. My research is about volcanoes and lava. The Messiah while on the cross suffered a wound to his right side from a spear. On September 10,01 I had a surgical incision on my left side of my body to repair a hernia condition. On Sept 10 almost 2000 years ago religious extremists plotted a heinous crime that affected the entire world. In the neighborhood where I live, there are numerous religious symbols, but they are not related to any Church or shrine. Near my home is the Ark, a non-profit organization. We remember the Ark of the Covenant and Noah's Ark. There is a pyramid home next to the Ark, for this I remember the great Pyramids of Giza. There is a street next to me called Iona, for this I remember the ancient Greeks and Ionia. Next to my home is a 70 ton granite T-rex stone creature that also looks like a giant serpent, this makes me remember the serpent where Jesus said "Be ye wise as serpent's." This also reminds me of Adam and Eve in the Garden and the resulting curse, which is what my research is about. This gigantic monolithic serpent also has a giant triangle-pyramid image in the front. There is a very popular water falls a short distance from my home and a little further on and up the mountain trail there is a beautiful place that locals call "The Garden of Eden.' The back of my yard is strattled by another county named Teller county. Edward Teller is the guy who invented the hydrogen bomb, which threatens Armageddon, which is what my research is about. I am not saying that I am the Messiah" there are just alot of weird strange realities in my life that make me uncomfortable. For instance, in Colorado there has been a Roman era. As a native of Colorado, I have lived under Love, Lamm, and Roman times. Like the Messiah who lived under brutal Roman times, I have lived under Governors John A. Love, Dick Lamm, and Roy Romer. This I find interesting.

The Location Of The 6 Electromagnetic Pyramid Images All Over The Earth

When we were children we had coloring and game books, in which we connected the numbers to form a "hidden" shape or we peered intently at a jungle scene that had a great amount of tangled vines and trees; as we tried to find the monkey or the other animals and objects, that were hidden in the thick maze. When we drew the lines from the preceding numbers, we were delighted when the connected number lines began to form the shape of a dog, cow or cat. We also felt a sense of accomplishment when we were able to locate the hidden animals or objects, in the dense jungle scene. Connecting man's ancient temple sites and the volcanic and earthquake zones around the world, is no different.

When we take a globe of the earth, we can plot the location of the 6 electromagnetic pyramid images all over the earth. With the help of an encyclopedia, showing the major volcanic regions of the earth; and man's ancient temple sites all over the world the six electromagnetic pyramid images begin to reveal themselves to us.

The location of the first electromagnetic pyramid image starts at the great temple sites of the Giza plateau in Egypt. This invisible pyramid image is roughly 500 to 700 miles wide, as it crosses over the entire earth. At the Giza area, there stands the greatest wonders of the known ancient world; which are the three great pyramids. This pyramid image is the most important one of the 6 pyramid images, because of the ancient Egyptian religious worship of Sirius and the universally recognized symbol of the pyramid—which is the calling card of the superior beings from Sirius. From the Giza pyramids we draw a straight line across the world to the ancient volcanic temple sites of the Aztec's in Mexico City. From the temple and volcanic sites of the Aztec's, we then draw another straight line down the ancient temple and volcanic sites of Central and South America until we come to the ancient temple and volcanic site of Machu Pich's Temple Of The Sun in Peru. From the volcanic mountain tops of the Inca's, we then draw another straight line across the earth back to the ancient pyramids at Giza Egypt. When we look at this astonishing pyramid image on the globe of the earth, we can see that it travels up from the Aztec empire through the volcanic fields of Mexico and then continues up into the south west and then the midwest sections of the United States. The left side of the pyramid image travels through New Madrid Missouri, the site of one of the most powerful and mysterious earthquake zones known to modern man. This fault in the earth is known as the New Madrid fault and it's origin defies all known geologic explanations, for the formation of earthquake fault zones around the earth. From this location the great invisible electromagnetic pyramid image then continues to travel across the face of the earth until it comes to the Great Lakes region; where the greatest concentration of people, ships, and airplanes have mysteriously disappeared.

This area was immortalized by the singer Gordon Lightfoot and the song that he recorded called "The Edmund Fitzgerald." He sang about the mysterious disappearance of this large ship in the Great Lakes region, From this location the giant electromagnetic pyramid image then travels to Canada's Lake Champlain, and the famous Loch Ness lake monster called by the name of "Champ." From this location the pyramid image continues on to the snows of Labrador and the site of a tragic 1985 jet liner crash that killed 200 U.S. servicemen; while on their way home from Egypt as well as the sinking of the Titanic. The pyramid image then travels past the volcanic lands of Iceland and then continues to Scotland and England, with it's ancient Stonehenge and numerous Roman temple sites; as well as Druid sites. From here the pyramid image keeps traveling past France, Germany, Italy, Yugoslavia, as well as all of the ancient sacred temple sites of Greece, Minos and all of the ancient temple sites of the Mediterranean and the lower Levant. All of these areas of the world are known for their volcanic and earthquake activity, which is tied to the volcanic basalt Rosetta stone. The second greatest of the 6 electromagnetic pyramid images is located in the Pacific Ocean, west of the United States. This pyramid image is actually as large as the 1st pyramid image, but is not as important as the first; because it does not intersect the great pyramids at Giza. This pyramid image is roughly 700 to 900 miles wide. It starts at the volcanoes of the Kamchatka peninsula of Siberia, site of a tragic jet airliner downing, by Russian air force jets in the 1980's. This event occured after the auto pilot had electronically caused the huge jet liner to stray into sensitive air space, over a Russian military complex. For reasons still unexplained, the electronics that control of the auto pilot had been affected by an unknown source. This caused the airliner to veer off course and fly over a restricted area.

Subsequently a number of years later, the Russian pilots involved in the tragedy as well as the Russian air force personnel, who were manning the electronic computer radar at the Kamchatkaall reported that their electronics and some of their radar stations, were not working for unexplained reasons. These events occurred at the time that the Japanese KAL jetliner was flying near the Kamchatka area, and it's sensitive electromagnetic computer systems suddenly malfunctioned. The airliner was then shot down by the Russian air force planes, over the Pacific waters. From the Kamchatka volcanoes the second greatest pyramid image continues on to the volcanoes of Japan and the Philippines, and then onto Indonesia and the areas of Krakatoa and Tambora. This is the site of the world's greatest volcanic eruption in the last 7,000 years. This gigantic volcanic eruption affected the entire earth, by eliminating the summer seasons in the northern hemisphere. This huge eruption occurred in 1815, and the date as well as the time frame of this eruption, is very significant. At this moment we must briefly return to the Rosetta stone and a very obscure ancient Egyptian sacred text called The Virgin Of The World. Most past and modern western classical scholars hate this Egyptian treastie because they have labeled it as neoplatonist and thus is a mystical work. The Virgin Of The World is a text that tells us that modern man will discover the true history of mankind, the earth and the universe. These events will occur through a series of discoveries related to the dawn of the space age.

The Rosetta stone was discovered by French engineers in 1799 and in 1822, just seven years after the huge volcanic eruption at Tambora in 1815, the world s greatest linguist named Jean-Francois Champollion, broke the silence of the ancient Egyptian hieroglyphics after thousands of years of mystery. He was the first linguist to translate the mysterious symbols, seven years after the monumental eruption of Tambora.

Mankind's ancient ancestors including the Egyptians all considered the numbers 2,3,5,7 and 9 to be magical, powerful, lucky and sacred numbers. The num ber 2 stood for the planet Venus, which modern American spacecraft have discovered, has the greatest concentration of volcanoes in our solar system. The number 3 stood for the earth and it's numerous world wide volcanoes, and lakes of volcanic fire. Subsequently the number 5 stood for the planet Jupiter and it's moons. Jupiter's moon Io was the most important because of it's profuse volcanoes and lakes of fire. Io is a moon that is completely consumed with seething volcanoes and oceans of molten magma. The number 7 was considered powerful and lucky by our ancient ancestors, because it represented the pyramid image and it was representative of ancient Egyptian Seven Shinning Ones. These 7 powerful Shinning beings were figures in the Egyptian religious worship of Sirius. The number 9 stood for the nine great Gods of the Egyptian Great Eneid. The number 9 also represented the nine planets of our solar system as well as the nine months gestation period of the human fetus, and the nine light years that Sirius is from the Earth. The number 9 was considered the sacred number of the Gods.

We can therefore understand that when Champollion broke the linguistic code of the ancient Egyptian hieroglyphics, seven years after the greatest Volcanic explosion at Tambora; The Virgin Of The World and it's prophesy that mankind would eventually discover the true history of man and the universe was now starting to come true. It was because of the fact that Champollion was able to read the ancient Egyptian writings, that mankind today was able to understand their religious worship of Sirius.

Returning to the second greatest world wide pyramid image at Tambora, the great pyramid image then crosses through Australia and then continues on past the volcanoes of New Guinea and the south sea volcanic islands until it reaches the pyramid shaped volcanic Easter Island. Easter Island which is the location of the mysterious volcanic megalithic stone statues, is anchored by three volcanoes and is connected to the ancient religious worship of Sirius These volcanic stone statues face inward, and their facial and physical features are very similar to those of the Pharaoh Akhenaton of the Egyptian Amarna revolution. Akhenaton is known to history as the heretic king in that during his reign, he abandoned the Egyptian pantheon of Gods, and instituted the worship of Ra the Sun God above all others. There is consensus among some scholars that Akhenaton laid the eventual foundation for the rise of Christianity, through the worship of only one God. It is no accident that this pyramid shaped volcanic island was discovered on Easter, which is one of the holiest days of the Christian religion.

From Easter Island the great pyramid image then travels up the Pacific ocean towards the Hawaiian Islands. Before it reaches these islands though, the giant invisible electromagnetic pyramid image travels through and past the location of the newly discovered; greatest concentration of under water active volcanoes in the world. These volcanoes were discovered in 1993. From these large submerged volcanoes, the pyramid image then continues to travel to the Hawaiian Islands home of the angry volcanic god Pelee. From this location the giant pyramid image travels further up into the Pacific ocean until it reaches the volcanoes of the Kamchtka peninsula.

The third world wide electromagnet is pyramid image is located in Europe.   It begins with the volcanoes and ancient Roman temple sites in Italy, and the famous Pompeii volcano. This large pyramid image then travels through the southern Mediterranean and it's volcanic regions of Greece, Turkey, Iran, and then onto India. The great pyramid image goes to the Himalaya Mountains, land of the Abominable Snow Man. It then turns north and crosses central Asia and travels onto Siberia where it meets the great volcanoes of the Aleutian Island chain of Alaska. From the volcanoes of the Aleutian Islands, the great pyramid image then changes direction and heads back to Siberia; and past the Arctic circle. At the top of the world.

The pyramid image then travels down the the earth past Norway, Sweden, Denmark, Germany, and then back to the volcanoes and ancient sacred temple sites of the Roman empire. This electromagnetic pyramid image is roughly 500 miles wide.

The fourth greatest pyramid image begins in the volcanic regions of equatorial Guinea Africa, and travels south through the south Atlantic until it reaches the volcanic islands of Tristan Da Cuna. From these volcanic islands, the pyramid image then travels to the southern boundaries of Africa, and then continues onto the Indian ocean and the island of Madagascar and the volcanoes of Mauritius. The gigantic electromagnetic pyramid image then travels back to Madagascar, and it's volcanoes and past the volcanic islands of Mayotte off the Madagascar coast. The pyramid image then enters the African coast at Tanzania and it travels onto the volcanoes of Rwanda. From Rwanda it travels past Zaire and onto the volcanoes and volcanic islands of equatorial Guinea Africa. The fifth and smallest of the five giant electromagnetic pyramid images is located largely in the United States. The pyramid image begins in British Columbia and Washington state, and the famous volcanoes of Mt.Ranier, Mt.St Helens, as well as other volcanoes of the Pacific north west. From these great volcanoes the pyramid image then travels south into Oregon and then into California, with it's volcanoes such as Mt. Lassen. The large electromagnetic pyramid image then continues to travel south to Mexico's California peninsula and the large volcanoes of this region. From here the pyramid image then turns northeast and crosses the peninsula waters and enters the volcanic regions and hot springs of western Mexico. From this volcanic region, the electromagnetic pyramid image then continues to travel up into north eastern Mexico, until it reaches the United States at the New Mexico and Arizona border. From this location of the country, the pyramid image then travels into New Mexico and past the great white sands of this region bringing to mind the bright Egyptian sands of the ancient Pharaoh's. From here the pyramid image then continues onto the middle of the state where it meets the volcanoes and lava flows of this region. From the volcanoes and black lava basalt flows of New Mexico, the electromagnetic pyramid image then reaches the state of Colorado at a point near Alamosa Colorado. From this region of hot springs and volcanic vents, which local residents as well as college students take advantage of there is also located the Great Sand Dunes. It is a mystery as to why the Great Sand Dunes is located in the middle of large alpine mountains. When we examine the sands of these giant dunes, we find that the sands consist of billions of tiny particles, which are magnetic. The large electromagnetic pyramid image that passes right through the Great Sand Dunes, has a magnetic charge that keeps the dunes in place; by way of the billions of tiny black magnetic magnetite particles. This area of the world also brings to mind the great sands of the ancient Egyptians. From this region the invisible pyramid image then travels to the black volcanic basalt plugs, and volcanoes near Trinidad and Walsenburg Colorado. From here, the pyramid image continues on to a point north of La Junta Colorado, and then changes it's direction.

At this juncture we will briefly examine a number of very strange and my sterious events that have taken place in Colorado, which is the entire apex of the fifth electromagnetic pyramid image of the world. In 1990 an air force B1B bomber mysteriously crashed just a few miles north of La Junta Colorado, while on a low level practice bombing run. All members of the crew were killed. In 1991 a large jet airliner mysteriously crashed while on approach to the Colorado Springs airport. All of the passengers were killed in the unsolved mysterey. The Federal Aviation Administration exhaustively investigated this baffling crash and could find absolutely no answer as to why this large jet liner crashed. The plane was flying normally while on approach to the air port, when it suddenly turned to the right and plunged vertically into the ground. Two weeks later a small two passenger plane also plunged almost vertically into the ground as well, killing it's pilot. The two crash sites are separated by less than 15 miles. All of these mysterious tragedies took place inside the apex of the fifth invisible electromagnetic pyramid image in Colorado. Officially recognized mysterious magnetic disturbances do exist at the apex in the state of Colorado. The Federal Aviation Administration, The Department of Defense and The Department of Commerce print a yearly sectional aeronautical chart map for airplane pilots. The map shows the Colorado Springs areas, as well as the state of Colorado. The chart shows that at ground level in Manitou Springs Colorado, there is a mysterious magnetic disturbance of as much as 13 degrees, which includes the entire city of Manitou Springs as well as Colorado Springs. The mysterious magnetic disturbance in the Pikes Peak region can affect man's electronic devices such as computers and other hardware. The best way to describe this magnetic disturbance is to compare it to throwing a rock into a still pool of water, and then watching the water rings expand outwards from the point of impact.

Electromagnetic waves are emitting outwards from the magnetic disturbance in Manitou Springs, to surrounding areas of Colorado. Many times local computer systems from banks, newspaper publishers, city traffic computers, and even the computers of the North American Air Defense Command, located at Cheyenne Mountain, are affected by the magnetic disturbance that is located here in Colorado.

Many times the computer systems of businesses and government go "down," for no apparent reason, and this remains a mystery.

For instance the North American Air Defense Command reported that in the early 1980's, a series of computer malfunctions at their complex, had resulted in some very serious problems. Mysterious computer failures reported that Russian ICBM's were being launched at the United States, when in fact no such attack was in progress. Military personnel could not figure out what was affecting the computers. This information was not voluntarily submitted to the public, but was leaked to the press. The news media over the years has reported that the General Accounting Office has revealed that the newly installed computers at NORAD are also having mysterious problems as well, and it will take additional millions of dollars to fix.

Over the years the local newspapers have reported that the computer systems that operate the traffic light system of Colorado Springs, has never operated properly. For reasons that are not understood, the computer system has never been able to work up to design specifications. To understand this problem, all one has to do is drive the streets of down town Colorado Springs.

The apex of the fifth giant electromagnetic pyramid image in the United States, is located in the whole state of Colorado. This brings up the location of the Garden of the Gods in Colorado Springs. For thousands of years the ancient indigenous Indian peoples of this area, have been worshipping at the Garden of the Gods. They worshipped mother earth, father sky and sacred mountain, in the Garden for millennia. At this sacred Indian religious location in Colorado, there exists five great natural megalithic edifices; which correspond to the five civilizations of mankind as spoken of by our ancient ancestors. The five natural rock edifices also are associated with the sacred number 5, which is related to the planet Jupiter; and it's red volcanic moon Io. Four of these rock megaliths are composed of red sandstone while the fifth edifice, which is located at the entrance of the Garden is composed of a rock called alabaster. When we look at this "soapy" stone, and we return to the ancient Egyptian pyramids; we discover that the ancient Egyptians carved many of their Sirius Gods from alabaster stone. Today many of these carvings can be seen in Egypt. In examining the stone sarcophagus of the ancient pharaoh, we find that many of them were carved from solid granite. Earth scientists as well as geologists tell us that granite is a volcanic or igneous based rock material. When we look up to the sky above the Garden of the Gods the towering Pikes Peak granite looms over us. This mountain is a volcanic mountain, but it is not the classic smoking mountain such as Mt. St.Helens. Pikes Peak however was formed from molten volcanic magma under the earth, which slowly cooled over time. The point is that the ancient indigenous Indian peoples of this region worshipped mother earth, father sky and sacred mountain, from the Garden of the Gods. When white settlers came to the Pikes Peak region, this sacred Indian religious site was confiscated and was renamed the Garden of the Gods. Everyone knew the ancient Indians had worshipped their Gods here, for eons of time.

Now we can return to the final leg of the fifth giant electromagnetic pyramid image, in the United States. From a point north of La Junta Colorado, the great pyramid image then turns north west and travels through Colorado. And arrives at Greeley Colorado, which is one of the nation's large bull and cow processing centers. The pyramid image then continues on towards the volcanic diamond kimberlite fields between Colorado and Wyoming. This is a volcanic area that straddles the Colorado and Wyoming borders between Fort Collins and Laramie; which is known as the State Line District. From this location the giant electromagnetic pyramid image then continues to travel into Wyoming, with it's volcanic hot springs in the state, and Devil's Tower on it's right boundary. The pyramid image then reaches the famous Yellowstone National Park with it's Old Faithful Geyser. From this volcanic area of the state, the pyramid image then enters the state; of Idaho with it's Craters of the Moon national monument. Throughout this area. there are giant volcanic lava flows over large areas. Further south west is located the Great Salt Lake, bringing to memory the sacred aura attributed to salt by the ancient Egyptians; who used natron (salt) to prepare the dead Pharaoh for embalming. From the volcanic regions of Idaho the pyramid image then travels back into Washington state and arrives at the volcanoes of Mt. St. Helens, Mt.Adams, Mt.Ranier, Glacier Peak, Mt.Baker, and Mt. Garibaldi of British Columbia. Thus we have the location of the previously undiscovered electromagnetic pyramid images of the earth.

Pyramid # 1 location Giza Pyramids to Aztec Empire of Mexico City. The location of this pyramid image correctly identifies the North and the South, during the American Civil War of 1860. Giza to Mexico City to Machu Pichu to Giza pyramids.

This pyramid image travels across the Earth to Giza Egypt and as it approaches western Europe it is intersected by the # 3 pyramid image, that covers Russia.

Pyramid # 2 location: The location of this pyramid image starts at the Siberian peninsula of Kamchatka and it's majestic mountain volcanoes. From here the pyramid image then travels to the volcanoes of Japan and the Philippines, and then to Tambora and Krakatoa. From this region the pyramid image then changes direction and crosses the south Pacific Ocean until it comes to the volcanic pyramid shaped Easter Island. From here it then travels back north until it reaches the volcanoes of Hawaii, and then returns to Kamchatka.

Pyramid # 3 location: The location of this pyramid is from the volcanoes of Italy, then through the southern Mediterranean to Iran and India. Then onto the Aleutinan Islands and then back to the ancient Roman temple sites and volcanoes of Italy. This pyramid image and it's left corner intersects near the left top of the # 1 pyramid image to create a 6th pyramid image.

Pyramid # 4 location:

The location of this pyramid image is situated in equatorial Guinea Africa. then onto the southern Atlantic Ocean and Tristan Da Cuna Island. From here the pyramid image travels up to the Mauritius From the Mauritius Islands, the pyramid, image travels back to the volcanoes of equatorial Guinea Africa.

Pyramid # 5 location: The location of this pyramid image is in the United States. It starts from the volcanoes of British Columbia and Washington state. From Mt. Ranier and Mt. St. Helens, the pyramid image travels south until it reaches the majestic. Mt. Lassen volcano. From this location the pyramid image then continues to travel south until it reaches the volcanoes of the Baja, next to the Gulf of California. From this location the pyramid image then turns north-east and travels to the volcanoes of New Mexico and Arizona and Colorado. From here the pyramid image then goes to Wyoming and Utah, and then back to the volcanoes of Washington state.

Pyramid # 6 location: The location of this pyramid is found in western Europe, and it touches Italy Germany, Austria, Yugoslavia and other countries in this part of the world. The base faces right at Japan. This pyramid image correctly identifies the instigators of World War I and II, and currently identifies Yugoslavia or Bosnia Hercegovina as the killing grounds. This pyramid image correctly identifies the Axis Powers of WW II and the # 6 verifies these wars as evil.

Pyramid #6 and it's meanings: This pyramid and its location which was the site of the greatest human carnage in history, has other meanings when it comes to the ancient sacred numbers. Mars is the 4th planet from the sun, and was the God of war, and destruction. The ancient Sirius religion tells humans not to kill other human beings. The #6 pyramid image is created when the #1 and #3 pyramid images intersect in Europe. When we add these two pyramid numbers together we get the number 4, which is the number of Mars, the God of War; further confirming this pyramid and the events that happened there as evil.

Pyramid # 1 and it's significance to the Nostradamus prophesy. This pyramid image covers the United States and it is intersected by the 6th evil pyramid. The # 3 pyramid image which covers Russia, is also part of the Nostradamus prophesy, when he says that the two great leaders will be friends. The United States and the Russian leader are now in fact talking to each other, thus confirming the first part of the Nostradamus prophesy.

The Scientific Detection Of The Six, Pyramid Images

There is strong evidence that the six electromagnetic pyramid images have already been detected by man's scientific instruments. Recent scientific investigations into gravity research by scientists, has begun to detect a mysterious fifth force on the Earth and in various parts of the world. In the May issue of National Geographic magazine, scientists have detected a mysterious new. electromagnetic force. There are four recognized forces in the universe which are the following:light, electromagnetism, gravity, and magnetism. Some science investigators from around the world have detected a mysterious fifth force, and yet other researchers have not been able to detect the new fifth force.

Sir Issac Newton discovered gravity, and his written discoveries and research into gravity and other scientific disciplines are considered by modern scientists to be absolutely without question. Scientists consider it heresy to question the works of Newton, in the field of gravity. It is exactly because of the giant stature of Newton, that the detection of a new fifth force in the universe is so unsettling, to the scientists who say they have detected this new force. Newton's laws of gravitation are universally accepted by scientists even in the face of a fifth force discovery. The prevailing thoughts among the new research scientists who have detected this new fifth force, is that they are upsetting the hallowed realms of Newton; who is the stalwart symbol of the laws of gravity. There is no doubt that Newton was a most brilliant man, and it is well known fact that he discovered gravity and authored many brilliant papers in his scientific endeavors. He also was reportedly a member of a secret organization of people, who claimed to possess great secrets.

The question that we must ask ourselves is, What is this elusive fifth force that does in fact represent the number 5 and which is the ancient sacred number for Jupiter To further delve into the realm of gravity and the legendary world of Sir Issac Newton, we must consult the May 1989 issue of the National Geographic magazine on page 563-83

The first researchers to discover the fifth force was Ephraim Fishbach. Fishbach a physicist and his fellow colleague physicist Samuel Aronson conducted their first fifth force experiments at Purdue University in Indiana. The two physicists began their experiments in late 1985. They were both studying a series of findings from an atomic accelerator or atom smasher. They discovered a set of results that they could not explain. Particles called kaons were behaving in defiance of gravity, inside the accelerator. The two researchers thought of every possible explanation for the gravity defying kaons, to no avail. They felt the only explanation for the bizarre observations was a fifth force in the universe. Purdue University is located in Indiana, where the two physicists conducted their fifth force experiments which did in fact detect a fifth force. The number one electromagnetic pyramid image travels right through Indiana, and it's surrounding areas such as Chicago Illinois.

At the same time that Fishbach and Aronson were working on detecting the mysterious fifth force, a geophysicist named Frank Stacey and his fellow colleagues at the University of Queensland in Australia were also conducting their own fifth force experiments, deep inside an Australian mine in Queensland. Their experiments were successful, and they also detected a fifth force opposing gravity. Since then, other experiments in boreholes and mines elsewhere in Australia have confirmed the detection of a fifth force. Stacey now believes that the observed and detected fifth force discovery, must be taken seriously. The geophysicist and his colleagues conducted their fifth force experiments in Australia. This is the exact location in Queensland, where the number two electromagnetic pyramid image travels through while on it's way from Tambora and Indonesia, while on it's way to Easter Island.

Attempting to also detect the fifth force, British geophysicist Bob Edge and Keith Runcorn conducted a series of experiments at a reservoir in Wales. They measured the gravitational pull of water, as the reservoir was emptied and filled. They too detected a gravitational deviation of about 5% from the norm. In their experiments they came up with the 5% deviation, which again is the number 5, and is associated with the ancient Sirius religion. When we look at where the geophysicists conducted their experiments, we realize that they were in Wales; near Ireland and England where the enormous number one electromagnetic pyramid image travels through.

Next we examine the experiments of Paul Boynton and his colleagues at the University of Washington in Washington state. The Physicist fashioned weights of aluminum and beryllium, and suspended them next to a gigantic wall in the Cascade mountain range. Boynton and his colleagues also detected variations in their fifth force experiments. As we know by now, the fifth electromagnetic pyramid image is located in Washington state, because of all of it's volcanoes. Boynton and his colleagues did detect variations, but they were not sure of what they detected.

Geophysicist Donald Eckhardt of the U.S. Air Force geophysics laboratory in Bedford Massachusetts and his colleagues also conducted the fifth force experiments, when they ascended a 2,000 foot tall T.V. tower in North Carolina. Eckhardts group of researchers also were successful, when they detected a significant departure from normal gravity. Eckhardt and his fellow colleagues findings were of such a significant departure, that he said he was sure of his results. When we examine where these scientific researchers were at when they conducted their fifth force experiments, we realize that they were inside the enormous number one electromagnetic pyramid image; that completely covers the eastern and southern United States. It is this area that AT&T suffered the massive failure of their communications hardware, in the early 1990's.

There have been critics of the fifth proponents, such as the physicist Alvaro De Rujula who in 1986 said, quote: "In a few years this fifth force rubish will be gone." Two years later in 1988 he was not as quick to dismiss the fifth force discoveries. He said, quote: "In the absence of any two experiments with the same results, we can't really say anything scientific about it yet." Another critic Princeston's John Wheller a leading theorist said, quote: "I think the fifth force will prove to be a flash in the pan."

The physicist Alvaro De Rujula was quite interesting in the statements that he made on the fifth force question. He said that there was an absence of common results, in the various tests that were conducted around the world. Each test that was conducted around the fifth force locations, all showed varying degrees of departures from the expected norm of gravity. The largest departure from normal gravity was detected in North Carolina. These varying test results from around the world, which were conducted on the various electromagnetic pyramid images; unbeknown to all the various researchers, are the exact kind of varying test results we would expect to see because the pyramid images are of varying sizes and strength.

We are now aware of the fact that physicists and scientific investigators from around the world have been detecting some sort of strange and mysterious fifth force, that is opposed to gravity! This force carries a magnetic reaction to gravity, and every one of these research teams were able to detect this force. Unbeknown to them however, every one of these fifth force detection research, teams; were conducting their experiments on one of the electromagnetic pyramid images around the Earth. What they were detecting was the energy of the electromagnetic pyramid images.

As a result of the dawn of the space age, scientist have been able to detect fluctuations in gravity around the Earth by sensors aboard orbiting space craft. For instance, gravity is very strong in the Himalaya mountains and in central Africa. Gravity is noticeably weaker in the Hudson Bay region, and the Indian Ocean. In central Africa and the Himalaya mountains, two of the very enormous world wide electromagnetic pyramid images are located. However on the other hand, in the areas of the Hudson Bay and the Indian Ocean; where gravity is much weaker, there are no electromagnetic pyramid images. When we examine the data from orbiting satellites, we find that the earth bulges off the coast of east Africa. The Lageos and Geosat spacecraft have determined that there is a powerful Indian Ocean anamoly off the coast of Africa. This area includes Madagascar and the surrounding areas, where gravity is quite strong. The fourth largest electromagnetic pyramid image is located right at this area. This area of the world is also known as The Garcia Quadrangle and is known as an area where people, ships, and planes have mysteriously disappeared. Therefore we are able to see that certain areas of the Earth have strong gravitational properties, which are connected to the electromagnetic pyramid images of the Earth. Areas of the Earth where there are none of the six invisible pyramid images, on the other hand are known for weaker gravity.

Daniel Masias (Muh Sigh Uh)

The Magnetic Properties Of Volcanic

Lava And Magma

It is a scientific fact that geologists and scientists will confirm, that volcanic lava and magma have magnetic properties. For reasons that are not entirely understood, the volcanic material has magnetic properties which can cause magnetic disturbances. Another very interesting geologic fact that most people are not aware of, is that granite is also a magnetic rock and it is formed from magma under the earth. Granite in essence is a volcanic fire rock, that slowly cooled under the earth over long periods of time. When we look at the earth and it's majestic granite mountains all over the earth, we then realize that the granite mountain ranges of the earth and the smoking volcanoes all around us, were all born of the same source. That source was the fire hot magma under the earth. Granite mountains are not the classic smoking and rumbling behemoths, that we are all familiar with such as Pikes Peak and Mt. St. Helen's.

A very interesting fact about the planet Venus is that it has the greatest concentration of volcanic activity from the ancient past. The planet's atmosphere is in a run away greenhouse effect. Both the Earth and Venus have the same amount of carbon dioxide, but the Earths carbon dioxide is concentrated in it's oceans. The carbon dioxide of Venus is located in it's atmosphere as opposed to the Earth. Despite the fact that Venus has massive ancient lava flows all over it's surface, we would expect this planet to have powerful magnetic fields around itself; but it does not have the magnetic fields that we would expect!

The planet Mars is also a very mysterious world as well, for it too has profuse amounts of lava flows as well as volcanoes, that are quite visible from orbiting satellites. Up to this juncture I have been discussing the electromagnetic pyramid images that surround the Earth. Now we will briefly turn to Mars and examine this planet to see what we can discover, in light of the volcanoes of the earth. American space probes of the 1970's have revealed that there are a number of truly gigantic volcanic mountains on Mars. One of them is called Olympus Mons and it stands 88,000 feet tall! The top of this volcanic mountain is 50 miles long! This particular volcano dwarfs any volcano on the Earth or for that matter, in the entire solar system. What is more interesting than these incredible facts, is that there are three other volcanoes that are located near Olympus Mons. When we look at a U.S. geological survey base map of Mars, we can see that there are four very large volcanoes on the surface of Mars. When we draw a line between all of these volcanoes, we find that a triangle or pyramid image then emerges from the four volcanoes! Just as on the Earth, where there are six.

Invisible electromagnetic pyramid images are all over the Earth. There is one pyramid image on the surface of Mars! I honestly do not know if this pyramid-image is electromagnetic in origin. Earlier I mentioned that our ancient ancestors said that the numbers 2,3,5,7, and 9 were sacred numbers. In ancient mythology Mars was the God of war, and this planet is the fourth planet from the sun. The Earth is the third planet and it's sacred number is 3. The ancients reveal to us that the number 4 is not a sacred number, because it represents war and destruction. In the ancient religious worship of Sirius, mankind is encouraged to treat his fellow man with kindness and understanding; and murder is forbidden. Therefore we can understand why Mars and it's number 4, was not considered a sacred number. In any case there is no doubt that there are four gigantic volcanoes on the surface of Mars, that corresponds to it's number from the sun. We would expect Mars to have powerful electromagnetic fields around itself, but it does not and it is more like Venus.

The prevailing scientific thought is that our solar system and it's sun, were formed from one common nuclear source billions of years ago. A gigantic rotating saucer shaped cloud of solid and gaseous material, which was radioactive in nature; which formed 5 billion years ago. Out of this seething caldron of intense heat, light and radiation, there emerged our solar system through the bonding effects of gravity. Because scientists believe that our solar system was formed "from one common source," then we must assume correctly that the volcanic materials inside Mercury, Venus, Earth, Mars and the solid moons of the large gas planets; would all be surrounded by powerful magnetic fields. This is not the case, because only the Earth, Jupiter and to a lesser extent, Saturn; have powerful magnetic fields around themselves! Therefore any reasonable person can understand that something very strange is happening in our solar system. The question is, why don't the other solid bodies of our solar system possess powerful electromagnetic fields around themselves? After all, all of these planets were formed from the same source! My question can be answered if we go back to the original hypothesis of the scientists, and question their assumption. That our solar system was formed from a gigantic rotating mass of hot radioactive material. In order to properly address this question, we must first look at the physical features of the Earth, in ways that we have never thought of. We must look at the surface of our planet in a manner that has never occurred to us, but before we can do this we must first address another subject that will help us to try and understand; that maybe our solar system was not formed in the manner that scientists have envisioned it. Before we can do this we must first try to understand what our ancient ancestors believed in. Ancient peoples all over the Earth referred to mother Earth, father sky, and sacred mountains, as the three sacred deities of nature; that was bequeathed to mankind and all living things. Here we see the reference to the powerful number 3, which represented the Earth and the four pointed pyramid, as well as the number 7. With reference to the sacred mountain, of which the Earths mountain ranges are made of; we know that these granite edifices are of volcanic origin. Here again we cannot escape the fact that the presence of the superior beings from Sirius are every where on the Earth, and their presence on the planet is total and complete; which includes our own bodies.

Historical Events, Legends, Mysteries And Tragedies On The Pyramid Images Of The Earth

There are some very odd and strange events surrounding the locations of these electromagnetic pyramid around the earth. These events have to do with disasters, mass murders, tragic historical events and the severe weather conditions of the earth. There are two pyramid images that exist in the United States, one of which is the number one and the most important one; and the number five pyramid image. The number one pyramid image which has it's origins from the Aztec empire of Mexico to the Inca empire at Machu Pichu and then travels to the three pyramids at Giza is partly located in the United States. This may be the reason that Nostradamus declared that America and Russia are the focus of his prophesy.

From the Aztec location in Mexico City this gigantic pyramid image crosses into Texas and the location where an astonishing number of events have taken place over the years. These tragic events include the assassination of an American President. This President was responsible for mankind's first landing on another celestial body, and was foretold in the ancient Trismegistic Literature, of the Egyptians.

The gigantic electromagnetic pyramid image enters the United States from Mexico, into Texas north of Laredo. It then travels in a northeastern direction, through the entire state. When we start to investigate news events that have occurred in Texas, we learn that in the early part of the century and even before this time frame; when ancient indigenous Indian peoples inhabited this region, a mysterious white light has been viewed at night near Marfa Texas. the mysterious light can be viewed on occasion by Texas residents, who have made diligent efforts to locate and identify this strange light source. Scientific efforts have been conducted to learn more about this light source, but all efforts have been in vain. The gigantic electromagnetic pyramid image travels through the area of Marfa Texas, and is responsible for this light source.

When we investigate other events in Texas we learn that in the early part of the 1960's, a man was involved in the mass slaying of people from a tower at the University of Texas. In 1963 no American can forget the assassination of President John F. Kennedy in Dallas Texas; the exact location where the large electromagnetic pyramid travels through. We then realize that in 1992 a man in Killeen Texas, drove his truck into a restaurant cafeteria and killed many people with his guns. In Texas there have been a number of large jet airliner crashes that killed many people, in the Dallas-Fort Worth areas. They took place during huge electrical storms and were eventually blamed on wind shear. When we continue to check the news events of this state, we then remember the Branch Davidian sect of Waco Texas (we ain 't coming out.) The resulting fire killed many innocent people. All of these tragic events took place on the giant electromagnetic pyramid image that travels through the state of Texas.

From Texas the pyramid image then enters the state of Missouri and Oklahoma. In June of 1993 an Oklahoma man slaughtered his family in a mass slaying that shocked the state of Oklahoma. The state of Missouri is where President Harry Truman was from, and remembered for the World War to bomb the Japanese. He was the first human in the history of mankind to order the use of an atomic bomb on Hiroshima and Nagasaki Japan, confirming a prophesy of Japan as being the rising sun. The other prophetic statement was made by Harry Truman when he said "the buck stops here." He did not realize the prophesy of his words, and what the ancient Egyptian pyramid and single eye on the American one dollar bill meant. All of the symbols on the dollar bill are the symbols of the superior beings from Sirius. From the ancient Egyptian pyramid with it's single oval eye to the symbols of the oceans, sea's, rosettes and it's scroll work, all of these ancient symbols of our ancestors; denoted the superior beings from the star system Sirius. Western civilization adopted all of these symbolism because we are tied to the ancient Greeks, because of our democracy. However in copying these traditional symbols the real meaning behind them was not known. A modern example would be to sign a contract, and not read the fine print at the bottom of the document. That fine print at the bottom is the key to understanding, what these ancient symbols of the Greco-Roman classic style really meant. All of these symbols are tied to the oceans and waters of the sea's as well as fish and amphibians, which are the very calling cards of The Masters Of The Waters from Sirius.

Ancient Egyptian sacred texts speak about the fact that God was born out of nothing but empty matter, and that he gave birth to himself from nothing but dark matter; which exists in the universe. The Bible of the Egyptian worship of Sirius also informs us that the light and the energy of the stars, is the very material that their God is made of.

Through some very mysterious and complicated manner, which they do not reveal God is made of the nuclear energy that makes the stars burn hot.

The Nostradamus prophesy talks about the use of two nuclear weapons in the harbor, which is of course Hiroshima and Nagasaki Japan. It does not take a great deal of intelligence to realize that if God is made of, or somehow associated with nuclear energy then the use of nuclear weapons to kill fellow human beings could well be very prophetic for President Harry Truman.

Since Harry Truman made the decision to use nuclear weapons against the Japanese, and he is known for the phrase "The buck stops here" which does indeed have the symbols of The Masters Of The Waters; then indeed the responsibility for his fateful decision does stop at his desk. Therefore we can understand why Nostradamus said that the United States is the focus of his prophesy, because through the decision to use nuclear weapons the buck truly does rest at the feet of President Harry Truman. Ancient Egyptian sacred texts inform us that humans are judged according to what is in their heart, and the water symbols on the American one dollar bill are indeed the symbols of The Masters Of The Waters, from the star system Sirius.

Returning to the gigantic electromagnetic pyramid image in Missouri, we remember that in the mid 1980's, a man in central Missouri was implicated in the mass slaying of seven members of his family. Further investigations in Missouri reveals other mysteries as well. For instance Missouri is the site of one of the most mysterious geologic earthquake fault zones in the world. Geologists and scientists cannot explain why the New Madrid fault lies in New Madrid Missouri, and what manner of unknown geologic source gave birth to this earthquake zone. The New Madrid fault does not fit into any known standard definition of how geologically, an earthquake fault zone is created. Geologists tell us that earthquake faults are formed when two powerful plates of the earth begin to collide against one another.

In New Madrid Missouri, this very powerful fault zone is completely devoid of any colliding earth plates and as a result of this geologists are hard pressed and at a loss to explain what unknown geologic agent created this powerful earthquake fault zone. There is absolutely no doubt that scientists are very concerned about the New Madrid fault, because the last time that it moved in 1811-1812; it moved the earth in a manner that has never been experienced before, in the United States. The Missouri river was turned into a raging over flowing caldron, and the surrounding lands were literally pushed up into the air in a most terrifying manner. Compared to the San Francisco earthquake, the New Madrid earthquake was much more powerful in size; but it killed few people because the area in 1811 was sparsely populated. It is through New Madrid Missouri that the gigantic electromagnetic pyramid image travels right through.

At this juncture we will briefly go back to the ancient Egyptian Rosetta stone, in order to better understand a series of seemingly unrelated events around the world. These events are independent of one another. The Rosetta stone was discovered in 1799, a date the ancient sacred numbers of 7 and 9. A whole series of events took place from 1799 to 1832, that are all inter connected to one another. The New Madrid fault in Missouri produced three gigantic earthquakes in 1811-1812, and they were accompanied by the stench of sulfur; that is typical of an erupting volcano. The world's greatest volcanic eruption occurred at Tambora in 1815, or three years after the gigantic New Madrid earthquake took place in Missouri. Past news also shows that astronomical telescopic observations of the planet Jupiter, reveal that the great Red Spot or Eye of Jupiter disappeared from 1713 to 1831. This is the time frame that scholars and linguists around the world, were attempting to decipher mysterious Egyptian hieroglyphics on the Rosetta stone.

Jean Francois Champollion was the first linguist to understand the ancient Egyptian writings in 1822, and in 1832 he died. He died the year after the Red Spot or Eye of Jupiter, reappeared on the surface of the planet. All of these events were related to one another, for they showed the relationship of volcanoes, earthquakes, the Rosetta stone and Jupiter. All of these elements are part of the ancient Egyptian religious worship of Sirius.

Jupiter is a very pivotal element in understanding who we as human beings are, and the true history of the earth. I can qualify that statement by saying that absolutely astonishing events are taking place at the giant planet Jupiter, that relates specifically to the earth. The earth and the Jovian moon Io are the only known bodies of our solar system, which have active volcanoes, as well as lakes of volcanic fire as spoken of by the ancient Egyptian sacred texts. It is because of this fact that a shocking mystery has attracted the attention of scientists around the world. The earth and Jupiter are the only two planets that have powerful radiation and electromagnetic zones around them. Scientists describe both of these planets as virtual magnets. Later on I will address the astonishing events that are taking place on Jupiter, and how they are related to the earth.

The bottom line is that a whole series of events have taken place on the earth, from the time of the discovery of the volcanic Rosetta stone in 1799 to 1831 and 1832 when the great Red Spot or Eye of Jupiter reappeared, and Jean Francois Champollion died.

Returning to the New Madrid fault and the earthquake of 1811-12, we discover that after some further geologic investigations, there are other fault zones in the eastern United States and Canada, that are very much like the New Madrid fault zone in Missouri. Their origins are also geologically unknown, and are a mystery to geologists. When we look at the giant electromagnetic pyramid image, we can see that it travels right through these mysterious earthquake zones in the United States and Canada.

When we look at some of the astonishing events that have taken place in Texas, Oklahoma, and Missouri on the pyramid image, we must ask ourselves if these events and mysterious geologic features of the earth are just a coincidence or flukes. For an answer to our question, we next turn to Illinois and Ohio. These are two states where the pyramid image travels right through. The answer to this question is that the state of Illinois is the home of the mass murder Richard Speck, who killed 8 student nurses in the early 1960's, in Chicago. There are however other shocking events that have taken place in Chicago before Richard Speck emerged on the scene.

We remember that our ancient ancestors from the Minoans and the Egyptians all worshiped the Apis bull and cattle. The fact is that ancient peoples from all over the world have revered the bovine ruminants, in one way or another. Chicago Illinois has been for over 100 years the main meat processing center for cattle. It has only recently been replaced by Nebraska. In addition to being the main cattle processing center, we also remember another incredible tragedy that took place in Chicago. On October 8, 1871, the great Chicago fire was supposed to have been started by Mrs. O'Leary's cow. The resulting great fire burned for 27 hours and it burned down 17,000 building's and left thousands of people homeless. The city of Chicago was literally burned to the ground.

When we continue our investigations into Chicago, we discover that Chicago was the site of a momentous event in the early 1940's. The brilliant physicist Enrico Fermi was a world renowned Italian scientist, who was born in Rome Italy. (The site of one of the six electromagnetic pyramid images around the earth.) Enrico Fermi and his fellow colleagues were conducting nuclear research at the University of Chicago, for the United States government. This was under the auspices of the top secret Manhattan Project. On December 2, 1942, Fermi and his fellow researchers were the first scientists in the world to produce the world's first controlled atomic chain reaction. Later Fermi moved to Los Alamos New Mexico, and the work on the atomic bomb continued. (Another electromagnetic pyramid image site in the United States.)

When we look carefully at Chicago we remember that in the late 1970's Chicago's O'Hare International airport was the site of the worst airliner crash in U.S. aviation history. While taking off from the airport the giant plane turned to the right and crashed, killing over 200 passengers. None of these events are a coincidence, because there are more events that have taken place in this state. In the mid 1970's a homosexual mass murderer named John Gacy, was discovered. Then in 1992 the mass murder and cannibal Jeffrey Dahmer was apprehended by the police. when one of his intended victims was able to escape.

When one investigates the Ohio and Illinois states in the 18th century, we discover that both Presidents James Garfield and William McKinley were both murdered, and they were from Ohio. We then remember that President Abraham Lincoln was from Illinois and he too was murdered. When we check the location of the large electromagnetic pyramid image in the United States, we discover that it also travels past Memphis Tennessee, which was the site of the murder of Dr. Martin Luther King. This occurred in 1968. As we look at the globe of the earth, with the pyramid images drawn onto to it's surface we can observe that the huge pyramid image that travels through Texas, Missouri and the other southern and northern states are in fact the states that fought among themselves during the civil war. This is associated with President Lincoln and his stand against slavery, and is associated with the well known civil rights efforts of Dr. Martin Luther King. The point is that the momentous war between the north and the south, is correctly defined by the boundaries that this gigantic electromagnetic pyramid makes as it travels through these states. In effect it's north and south boundaries correctly identify the two combatants in this tragic war. From the Illinois area, the pyramid image then travels onto Michigan, and the site of a mass murder at a U.S. Post Office in 1990. From this location the pyramid image is now in the Great Lakes region of the world. There are 5 great lakes, which corresponds to the sacred number 5, and is associated with Jupiter. This is an area between the United States and Canada, where more people, airplanes, and ships have mysteriously disappeared than in any place in the world including the Bermuda Triangle. This also includes the Garcia Quadrangle, off the coast of Madagascar as well as the Bermuda Triangle off the coast of Japan. The singer Gordon Light foot sang a song about this mysterious area, as I have already mentioned.

From the Great Lakes region, the huge pyramid image then travels into Canada's Lake Champlain, and the location of the legendary lake monster sea serpent who is the cousin of the Loch Ness monster of Scotland. This legendary sea serpent is called Champ. From Lake Champlain, the pyramid image then continues to travel to Canada's north eastern provinces, and reaches the snows of Labrador. At this site in 1985 an American airliner with over 200 U.S. servicemen were killed when the plane mysteriously crashed, while on a flight from Egypt.

During World War II, a whole flight of American fighter planes also mysteriously crashed while on their way to England. From the mysterious and tragic plane crashes of this area, the invisible pyramid image then continues to travel out into the north Atlantic ocean. It is at this location that the famous Titanic ocean liner sank in 1910. The Titanic was carrying the mummified body of the Egyptian princess Taheb. From the Titanic site in the North Atlantic, the pyramid image then travels past the beautiful volcanic lands of Iceland, and then continues onto Scotland and it's legendary Loch Ness sea monster called Nessie. From Scotland the pyramid image then travels past Wales and Ireland, where for many years the Irish Republican Army has been waging a bombing campaign against England. As the pyramid image enters Great Britain, we then realize that it travels right past the ancient ruins of Stonehenge, as well as the many ancient Roman temple sites in Great Britain and the many other mysterious sites attributed to the Druids. In England after more research, we then discover that London was ravaged by The Great Fire Of London, much like Chicago was burned to the ground. The great bubonic plague also took place here as well as the famous mass murders of Jack The Ripper. In addition, England is well known for the profuse number of castle haunting's and spirit activities in this country. When we look at history of England we discover some very interesting facts that are related to dogs. The dog was bequeathed to man by the superior beings from Sirius, which is known as the Dog Star. Dogs can hear and tell when evil spirits are nearby, thus they bark at what appears to be nothing that can be seen. In fact there are ordinances all over the world, attempting to muzzle the senseless barking of dogs who are only doing what they are supposed to do for man. In England, the British Terrier is known as the most prolific dog barker in the world. This is no surprise because as we all know, Britain is famous for it's many spirit hauntings and it's ghosts. The British Terrier dog is merely doing his intended job of warning his owner of the presence of so many unseen evil spirits. When we look at modern England, we are reminded that this country has the greatest concentration of crop circles in the world as well as many documented UFO sightings. As the giant electromagnetic pyramid image travels through this part of the world, we are reminded of the tragic bombing of the 747 jet airliner over Lockerbie. When we begin to further investigate Great Britain, as pertaining to strange and mysterious events concerning human beings we uncover the fact that this country has the greatest concentration of human combustion cases. The other area of the world with large numbers of recorded human combustion cases is in the United States, in the Chicago and the mid west areas of the U.S..

From England the giant electromagnetic pyramid image then travels on to France, and Germany in tandem, and then through Austria, Yugoslavia and Greece. From the ancient temple sites of Greece and the Mediterranean, the pyramid image then returns to the three great pyramids at Giza.

Returning to the smaller or the fifth pyramid image in the United States, which is located from British Columbia and Washington State to the California peninsula; and then onto the volcanoes of Colorado we find that there are some interesting events that have taken place on the boundaries of this giant electromagnetic pyramid image. In checking past news accounts of the Pacific northwest we discover that in 1947 an air plane pilot was flying near the Mt. Ranier volcano, and he spotted a group of 9 flying disks near the volcano. Again this event underscored the many hints that mankind has been given, to help him to learn the true nature of his history. The flight of 9 silvery flying disks that the pilot spotted, was representative of the 9 great Gods of the Egyptian Ennead. This event was a watershed sighting, because it ushered in the modern term UFO and Flying saucers; that have been used ever since to describe these mysterious flying craft.

The specter of mass murder however has not escaped the fifth electromagnetic pyramid image in the Pacific northwest, because this was the home of Ted Bundy. Bundy began his killings in Washington State and continued them into Oregon, California, Utah, and Colorado. Every one of these states are located on the invisible pyramid image, here in the United States. The Green River mass killer who has never been caught, also began his killing's in the Pacific Northwest. The world's first nuclear processing plant is located in Washington State, at the Hanford processing plant in Richland Washington.

As the large electromagnetic pyramid image travels from Washington state south into Oregon, it reaches the many volcanic mountains of that state. From Oregon and it's volcanoes, the pyramid image then arrives in California, where there are large volcanoes and lava flows. One of these great volcanic mountains is Mt. Lassen. Mass murder in California has been very common, on the pyramid image. In the mid 1970's at Yuba City California, a mass murderer was involved in the killing of numerous migrant farm workers. We then remember that Robert F. Kennedy was assassinated at the Ambassador hotel in Los Angles. This event had historical implications, because he was running for the Presidency and the assassin was a man from the Middle East whose name was Sirhan Sirhan. Upon further investigations in California, we learn that there were a series of ghastly mass murders that were committed in this state. A man with an automatic weapon walked into a McDonald's restaurant and began shooting innocent children and adults. (After he told his wife that he was going hunting.) In a shopping mall in California, a mass murder also took place near Sacramento. Most recently a postal worker was involved in a mass slaying in that state. On July 1, 1993 a gunman walked into a

San Francisco highrise building and killed a total of 9 people including himself. All of these events took place on the electromagnetic pyramid image that travels through the western Pacific states. We then remember that in the early 1990's, the greatest fire disaster to strike the United States occured in southern California with the burning of a small grass fire. This fire resulted in a much larger property loss, than the great fire of Chicago in the 19th century.

As with other mysterious locations around the world, where the invisible pyramid image is located, the fifth pyramid image is also the home of numerous mass slayings. When we investigate the pyramid image boundaries between California and Mexico, we then remember that in the early 1980's a number of devil worshippers near the Mexican border was involved in the sacrificial mass murders of a number of victims in this area. As the electromagnetic pyramid image travels from Mexico into southern New Mexico, it enters the southern part of the state. The 300 hundred mile wide pyramid image travels into New Mexico near Deming and El Paso Texas, and then arrives at the lava beds and the White Sands National monument at the same time. Past newspaper accounts reveal that a number of astonishing events have occurred in this area of New Mexico. We discover that in the 1940's, the first atomic bomb test was conducted near Socorro New Mexico. This momentous event in human history occurred just 25 miles south east of Socorro in the area known as Dead Man's Route where the Trinity atomic bomb test was first conducted. Like the famous phrase that President Harry Truman coined, (the buck stops here,) the Trinity site called Jornada Del Muerto or Dead Man's Route, was a very prophetic phrase. When the two atomic bombs were dropped on the Japanese people, each bomb killed 70,000 people, Likewise the name for the atomic bomb test was called the Trinity site, which means, Father, Son, and Holy Ghost or three. The number three in this context represents the earth, and the implication of Trinity, is that there is a God like involvement here. This is exactly what the ancient Egyptian sacred texts, as well as the Nostradamus prophesy talks about in the context of the sacred white light of the Gods. God is somehow composed of the nuclear energy of the stars, therefore like all the other prophetic phrases throughout history the selecting of the name Trinity for the first atomic bomb test, was also very prophetic.

Taking a closer look at the Socorro area of New Mexico, we then learn that this area of New Mexico is also the site of the worlds most famous flying disk incident. In 1947 the news media reported that a crashed UFO with a number of alien occupants, was recovered by the U.S. government near Socorro. Years later, people who are still alive have come forward to say that they witnessed the recovery and saw the debris from the crashed UFO; near Socorro New Mexico. These witnesses have sworn that the events did in fact take place. The large electromagnetic pyramid image travels right through this area of the world.

From the volcanic lava flows of the Socorro area, the invisible pyramid image then continues to travel north east to the Los Alamos research complex of World War II. It was at the Los Alamos region that Enrico Fermi and Robert Oppenhimer along with many other researchers were able to develop the worlds first atomic bomb. This project was called the Manhattan Project and it was a top secret government endeavor, whose end results was spoken of by Nostradamus.

From the Los Alamos area and it's nuclear research, the electro magnetic pyramid image then continues into the state of Colorado and it's volcanic hot springs near Alamosa Colorado. As I have already mentioned, the pyramid image then continues past the black volcanic plugs and volcanoes of Trinidad and Walsenburg Colorado to a point north of La Junta Colorado. From this location the invisible pyramid then travels to the volcanic diamond fields of Colorado and Laramie Wyoming, and Yellowstone National Park. The pyramid then continues onto the lava and volcanic areas of Idaho and then onto the volcanoes of Washington State. The location of the fifth electromagnetic pyramid image in the western United States, and the direction that it points brings to mind a number of unsolved mysterious events that have, taken place in the United States. The mid west and western states have all experienced numerous cattle mutilations in the 1970's and 1980's. In the mid western states that grow wheat crops, a number of these states have experienced the formation of the mysterious crop circles. In the ancient Egyptian Bible or The Egyptian Book Of The Dead, much is written about wheat and wheat products such as cakes and other crop products. Likewise the ancient Dogon people also mention prominently wheat and barley, as well as other crop grains that are very important to the ancient worship of the star system Sirius. Just as wheat crops were very sacred to our ancient ancestors, the cow and the bull were also sacred as well. Recently cattle mutilations have begun to take place again in Colorado after a hiatus of 10 years.

Severe Climatology On The six Pyramid Images Of The Earth

There are severe weather factors that are associated with the 6 electromagnetic pyramid images of the earth. For the purposes of familiarity, I have concentrated on the pyramid images here in the United States. There are certain areas of the country that have some of the most severe weather patterns in the world. When we look at the base of the fifth pyramid image in the United States we can see that the Sierra Nevada mountains receives profuse amounts of snowfall during the winters. It is not uncommon to see 7 feet of snow on the roofs of this area. After a drought of 7 years in California, huge amounts of rainfall finally ended that states lack of precipitation; when there was extended rainfall and subsequent flooding in this state in the early 1990's.

When we look at the state of Colorado, which is the entire apex of the invisible pyramid image we know that this area of the world experiences some of the most concentrated electrical lightning storms during the summer months. This state is also famous among insurance companies and meteorologists, for the severe hail storms that pelt this state. In the late 1980's, Denver Colorado and surrounding areas, was devastated by the worst hail storm on record. The final bill to the insurance companies and policy holders, was a quarter of a billion dollars. In addition Colorado is ranked third in the nation for the formation of tornadoes, which is right behind Kansas and Texas. When we look at our map of the pyramid images in the United States, we see that the two pyramid images run parallel to one another. The number one, and the smallest pyramid image or number five are located right next to each other. From the Gulf of California and west central Mexico, these two invisible electromagnetic pyramid images run in tandem together all the way up into the United States to a point near Kansas, Texas, Oklahoma, Arkansas and somewhat near Missouri, and of course New Mexico and Colorado. These are the exact states that produce the most severe and greatest number of tornadoes in this country. Upon further investigations of this area of the world, we then realize that this area is the exact location of two famous areas. This area is called the Bible belt, and is the exact location of what meteorologists refer to as, "Tornado Alley."

When we begin to peer closer at the location of the gigantic pyramid image as it travels through the mid western states, we are able to recognize that this invisible electromagnet pyramid image completely covers the midwestern states of the Mississippi and Missouri rivers that have experienced the greatest flooding to occur in this region of the world. Weather maps and satellite views of this area, show the high and low stationary weather patterns, were lined up against the pyramid images; during all of this tragic flooding. If one watches the weather news, the various weather patterns can be seen to eddy or swirl around and follow the boundaries of the largest pyramid images as it travels from Mexico into Canada.

Because these two gigantic electromagnetic pyramid images are located right next to each other, in the middle of the United States they are the electromagnetic impetus for the various devastating storms that ravage this part of the world. The concept that I am conveying is that these two huge invisible pyramid images are providing the electrical energy, that fuels these storms. Under the earth, the volcanic material is magnetic in origin and is part of the fifth pyramid image in the United States. The combination of all these elements has thus come into play, and the results are the severe storms.

Further confirming the electromagnetic nature of the Earth and the number one pyramid image in the United States, we remember a number of highly publicized events that took place in the eastern United States. In this case we must remember that the entire northeast, as well as the southern United States is located inside the number one pyramid image. In the early 1990's the communications on the entire east coast went out. The communications giant AT&T on two separate occasions suffered huge company wide set backs, when their communications hardware completely failed. The resulting failure of their hardware resulted in the entire east coast, being subjected to a complete break down of computer and phone systems. Every aspect of human endeavor that involved some form of electronic communication was adversely affected by the break down. The communications systems of universities, colleges, businesses, corporations, government offices, airports and many thousands of other large computer systems as well as untold numbers of phone customers were all affected. All of these systems went "down" for reasons that were not explained. Because the earth is known to be a literal magnet, and the giant electromagnetic pyramid image completely covers this entire region of the world, there is little doubt that man's electronic devices can be adverse ly affected by the invisible pyramid images.

EMBLEMS OF THE HOLY TRINITY

THE TRIANGLE

The doctrine of the Trinity was manifested at the beginning of our Lord's ministry when the Holy Spirit descended upon Him at His baptism and at the same moment the voice of the Father was heard proclaiming Him as His beloved Son. (Luke 3, 22.) Again at the close of His earthly ministry, in His final instructions to the Apostles before His Ascension. The Blessed Saviour authenticates our belief in a Triune God. "Go ye therefore, and teach all nations, baptizing them in the name of the Father, and of the Son, and of the Holy Ghost." (Matt. 28, 19.)

This doctrine was taken for granted in the early church, but the first use of the word "Trinity" was at the beginning of the third century.(a)

The Early Christians shrank from making pictorial representations of so deep and sacred a mystery as the Trinity, but symbolic forms were developed later to use in defense of the doctrine when controversies arose within, as well as outside of the Church. There were very few of these emblems until after the 9th century.

The equilateral triangle is probably the first emblem of all, but early examples even of this are rare. It was, however, found in the catacombs. A very simple design, but it is most expressive.(b)

Two triangles forming a six-pointed star are also emblematic.

(a) Said to have been first used by Theophilus. Bishop of Antioch, or Tertullian, at the beginning of the 3rd century. (Webber, p. 39.) A translation from the Latin, of a hymn written by St. Ambrose, 340-397, begins:

"O Trinity of blessed light,
O unity of princely might:'
Hymn 11.

(b) "The use of mathematical symbols does not imply that the existence of God Is a matter for mathematical demonstration. Reason induces belief and faith aids man and confirms his conclusion." From an article by Agnes Peter.

COLOR: Green for Trinity season.

Daniel Masias (Muh Sigh Uh)

GOD

and the Bishop of Woolwich

God the Creator as depicted by William Blake

This is the tract on "Christian mythology"
that shook the Anglican Church
and announced a spreading revolution
in religious belief

Daniel Masias (Muh Sigh Uh)

EMBLEMS OF THE HOLY TRINITY

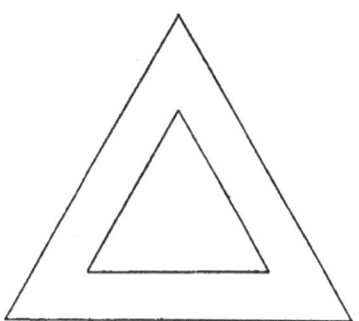

THE TRIANGLE

The doctrine of the Trinity was manifested at the beginning of our Lord's ministry when the Holy Spirit descended upon Him at His baptism and at the same moment the voice of the Father was heard proclaiming Him as His beloved Son. (Luke 3, 22.) Again at the close of His earthly ministry, in His final instructions to the Apostles before His Ascension. The Blessed Saviour authenticates our belief in a Triune God. "Go ye therefore, and teach all nations, baptizing them in the name of the Father, and of the Son, and of the Holy Ghost." (Matt. 28, 19.)

This doctrine was taken for granted in the early church, but the first use of the word "Trinity" was at the beginning of the third century.(a)

The Early Christians shrank from making pictorial representations of so deep and sacred a mystery as the Trinity, but symbolic forms were developed later to use in defense of the doctrine when controversies arose within, as well as outside of the Church. There were very few of these emblems until after the 9th century.

The equilateral triangle is probably the first emblem of all, but early examples even of this are rare. It was, however, found in the catacombs. A very simple design, but it is most expressive.(b)

Two triangles forming a six-pointed star are also emblematic.

(a) Said to have been first used by Theophilus, Bishop of Antioch, or Tertullian, at the beginning of the 3rd century. (Webber, p. 39.) A translation from the Latin, of a hymn written by St. Ambrose, 340–397, begins:

> "O Trinity of blessed light,
> O unity of princely might."
> Hymn 11.

(b) "The use of mathematical symbols does not imply that the existence of God is a matter for mathematical demonstration. Reason induces belief and faith aids man and confirms his conclusion." From an article by Agnes Peter.

COLOR: *Green for Trinity season.*

[23]

Daniel Masias (Muh Sigh Uh)

Published on January 7, 1998
Gazette Telegraph, Colorado Springs, Colorado

...and dispose of old tires. He'll

www.jobleads.org

a ministry of CRISTA

Serving Christian ministries since 1907

ATTENTION - Millenium 3000 DISCOVERY

THE MOST ASTONISHING WORLD WIDE (EGYPTIAN-GREEK-ASIAN-BIBLICAL) ARCHAEOLOGICAL DISCOVERY, IN THE HISTORY OF MANKIND - HAS BEEN UNEARTHED BY DANIEL A. MASIAS. AN INDEPENDENT EGYPTIAN-GREEK RESEARCHER, BUILDER & CARPENTER FROM GREEN MOUNTAIN FALLS, COLORADO. INTERNET ADDRESS IS: http://www.bytewarecafe.com:81

WHAT IS THIS DISCOVERY ABOUT AND HOW WAS IT UNCOVERED?

1 I looked at Mars (known to the ancients as the god of war and destruction) and found that 4 giant volcanoes there form a triangle.
2 From my research I knew that a number of ancient texts from the Egyptian Book of the Dead to the Bible, all mention geologic events such as lakes of fire, sulphur, smoking rumbling mountains and so forth. These geologic events are all associated with God
3 My Egyptian studies also revealed that many tombs including the Rosetta stone were composed and built of granite basalt, and diorite - which are earth materials that are all volcanic in origin.
4 From the moon landings I was also aware that volcanic hot spots were discovered on the moon. The 3 hot spots on the volcanic moon form a triangle like the triangle on volcanic Mars.
5 From my Bible studies I knew that Jesus was a Jew and the religious symbol of the Jews is a geometric configuration, of 2 intersecting triangles. The Star of David also has 6 additional triangles, that have in a clockwise 360° at the tips of the star. Because of the holocaust, I reasoned that the yellow star must have another meaning.
6 I also look into account the fact that the 15th century scientist Descartes proclaimed that Gods existence could be demonstrated in geometrical fashion. A triangle is a geometric symbol, the Star of David.
7 Finally in addition to other clues not mentioned here, I added the reversal knowledge about the Bermuda triangle. In this triangle, ships, planes and people mysteriously disappear and death is final.
8 Love triangle where 1 or 2 people are (history repeating itself)

WHAT DID I DO WITH THIS KNOWLEDGE?

I had an idea that I asked myself. What would happen if I took an Earth globe and made a red mark at all of the Earth's major volcanic earthquake and temple sites around the world. When I did this I was astonished to see that 6 triangles formed all over the Earth, just like the triangles on Mars and the moon. When the Bermuda triangle is added there are 7 triangles all over the Earth. Mankinds entire existence depends on 7 continents as well as the 7 days or 365 days of the year orbiting the sun. The #7 is also associated with all sorts of religious themes and is considered the ultimate lucky number. The #7 is actually a 2 sided triangle with one side missing. Mankinds brain and soul which distinguishes him from animals, is manifested in the skull. The skull has 7 openings - 2 ears - 2 eyes - 2 nostrils - 1 mouth. Man s brain makes war possible, relating to the 7 triangles

WHAT IS THIS ARCHAEOLOGICAL DISCOVERY ABOUT?

I have discovered that there are 6 more Bermuda Triangle images that exist all around the earth. On the boundaries of these 6 ancient triangle images, mankinds terrible wars, genocides, plagues, disasters and natural calamities have been taking place for many thousands of years. Just as death takes place in the Bermuda Triangle, death also occurs on these 6 mysterious triangles. *Source: (Genesis 2:8-17)*. Major Bible Prophecies, John F. Walvoord 1994 - Pg 23 25

WHERE ARE THESE MYSTERIOUS TRIANGLES LOCATED ON EARTH?

(Note: take a red magic marker and a globe of the earth and draw a red dot at all of the world's major volcanic, earthquake and temple sites around the world) The following locations are those sites.
Triangle #1: Giza Egypt to Aztec Mexico City to Inca Peru
Triangle #2: Kamchatka peninsula to Tambora Indonesia to Easter Island
Triangle #3: Italy to Afghanistan to the Aleutian Islands.
Triangle #4: Equatorial Guinea Africa to Tristan Da Cuna to the Mauritius Islands near Madagascar
Triangle #5: Colorado to Washington State to Baja California (La Paz) Colorado Volcanoes, Hot Springs, Sand Creek Massacre, Big Thompson Flood, 737 Crash, Suicide, Hail, Albuquerque Lows - Wyoming - Montana Yellowstone Volcanic Activity, Cheyenne, Sioux Indian Wars, Custer s 7th Cavalry Battle. Idaho, Craters of the Moon National Monument, Volcanoes, Hot Springs. Washington, Mt. St. Helens Eruption 5/18/80, Mt. Rainier, Teton Nuclear Pollution, Oregon, Volcanoes, Mt. Hood, 3 Sisters Volcanoes, Green River Killer, California, Volcanoes, Mt. Shasta, Mt. Lassen, Mammoth Lakes, San Francisco Great Earthquake, Mass Killers, Plane Crashes, Great Fires, Dam Rape California, Volcanoes, Mysterious Disappearance of Ancient Native Peoples. N.W. Mexico, Volcanoes, Poncho Villa Revolution. New Mexico & Arizona, Volcanoes, Wyatt Earp, Tombstone, Bord Hill, Geronimo, Cochise, Apache Wars, Billy the Kid, Colorado Great Sand Dunes, Ancient Egypt, Evangelical Christian U.S.C.G.
Triangle #6: This triangle is formed in western Europe by the intersection of triangles #1 and 3 which equals 4. Mars is the 4th planet and the god of war. The 6th triangle correctly identifies man s most horrible holocaust and terrible wars, genocides and plagues.
Triangle #7: The Bermuda Triangle, off the coast of Florida. *Source: Grollers Encyclopedia, the New Book of Knowledge, 1970 Vol. 19 P385 (NOTE. History repeats itself because of these triangles)*

WHY IS THIS DISCOVERY AN EGYPTIAN-GREEK-BIBLICAL DISCOVERY?

What I have discovered is the astonishing secrets of the Egyptian God Osiris and the forbidden knowledge of Genesis. I have uncovered the sacred knowledge that Adam and Eve learned about when they ate the apple (Wash. St. volcanic apples) from the tree of knowledge. I have been able to discover the physical (invisible) locations of the 7 triangle images of the earth. God used these cosmic triangle symbols when He pronounced a curse on mankind and the entire created world. He activated these images because Adam and I've disobeyed his command. The 7 deadly sins of the Christian Bible which mankind is cursed by, continues to energize these triangles and that is why there are wars, genocides, plagues, catastrophes, and natural disasters. In the Virgin of the World, Osiris is said to have buried mysterious cosmic symbols around the world. These mysterious elements of the cosmos were said to be buried and hidden in zones of the earth (earthquakes and volcanic zones.)

WHO IS DANIEL A. MASIAS?

Daniel A. Masias was born in Colorado and has been involved with his Egyptian-Greek research from primary Catholic school to his days at the University of Colorado. Daniel has a very strange birth-right He was born with many Egyptian, Greek, Asian, Biblical and Astronomical coincidences and common traits.

WHAT ARE THESE EGYPTIAN-GREEK AND ASTRONOMICAL COINCIDENCES?

1 Daniel was born on May 18, 1949 (Pope John Paul was born on May 18th) *Source: Olero County, La Junta Colorado Birth Records, Dr. Hiroshima, Gazette Newspaper 3/14/96*
2 The ancient Chinese consider May 18th to be a special lucky day *Source: Gazette Newspaper story May 23, 1996*
3 On May 18, 1980 Mt. St. Helens erupted (Helen is my mother's name whose father was a strong hoarder in Craig, Colorado) I own the copyright to the book I published that reveals the discovery of the 6 volcanic triangle images of the earth *Source: National Geographic, Jan. 1981 P 4*
4 Daniel has the same last name as Pharaoh Amasia who Herodotus called the great Pharaoh in 540 B.C. 26th dynasty. *Source: Pella to the Ancient Past, Tom B. Jones. 1967 P.86*
5 Daniel's last name Masias is Egyptian and Greek in origin. *Source: The Bible As History, Werner Keller, 1981 P 172*
6 The ancient Greeks believed that a messenger from the gods would appear in the west and his name is Messias *Source: 20 Volume Oxford English Dictionary, M for Messiah P 660*
7 Some modern Jews believe that the Messiah was born in 1949 when Israel was born. Daniel was born on May 18, 1949 and he lives only a few miles from Victor, Colorado where Jews from Denver want to build a 3rd temple to the Messiah *Source: Gazette Newspaper Dec. 11, 1993 Dec. 21, 1993 (NOTE. Moses, Ahmose, Amasis means in Bring, Remove, Extract or Take Out.)*
8 Asia is in Daniel s last name Masias *Source: Gazette story Dec. 21, 1997. Daniel discovered disaster triangles, Asia is the source of deadly flu.
9 In 1996 a red cow (Taurus) was born in Israel and the birth of this cow is associated with the arrival of the Messiah. *Source: Newsweek, May 18, 1997*
10 Mala (Maylas) was the mother of Hermes, he is the messenger of the gods. Mara was the eldest of the 7 Pleades stars. The number 7 is associated with the gods and Daniel discovered the 7 triangle images of the earth. Daniel's auto license plate is KNM 777 *Source: Websters Dictionary M for Mara*
11 Daniel's discovery is associated with the Sirius and Orion star systems 5 Egyptian female gods had star speckles on their backs, which denoted the abode of the gods. Daniel was born with speckles on his shoulders that show the shape of the Sirius and Orion star systems *Source: Encyclopedia Britannica M for Mythology P 976 Penruse Library*
12 Biblical Daniel 12.10 (The wicked will continue to be wicked. None of the wicked will understand. But those who are wise will understand.)

Daniel realizes these statements are different, but this is the way he was born. Every one of these statements can be verified with legal documents. Daniel is available for speaking engagements at a nominal fee. To learn more details about this discovery access my internet address. I am fighting the same ignorance that Galileo, Copernicus and others fought. This discovery helps to expand all the senseless violence on Earth! If mankind can learn about this discovery all men will be brothers and can live in true peace. Tax this story to other newspapers and friends around the world. Please help me spread this message of peace as I have no one to help me. Don't send me money. Please fax this discovery to other newspapers for publication Daniel A. Masias. P.O. Box 14. Gr. Mtn. Falls, Colorado 80819 Ph. 719 684 9169

Saturday, September 8, 2001   THE GAZETTE

THE WEST

Megavolcano simmers under park

Scientists see signs of life in Yellowstone caldera they call 'The Beast'

By Usha Lee McFarling
Los Angeles Times

YELLOWSTONE NATIONAL PARK — When the volcano here blew, it obliterated a mountain range, felled herds of prehistoric camels hundreds of miles away and left a smoking hole in the ground the size of the Los Angeles Basin.

But there are no signs in modern Yellowstone, aside from bubbling mud pots and geysers, that visitors are wandering through the caldera of one of the largest active volcanoes in the world.

"This is a geologic park, and not many know it," said Robert Smith, a geophysicist at the University of Utah who has spent his career piecing together the story of the Yellowstone volcano.

New sensors have allowed researchers to confirm a suspicion that Smith has held for a long time — that the ancient volcano scientists dub "The Beast" is a living force. The instruments record a continuing pattern of heaving and bulging and act as an early warning system.

Installed without fanfare and hidden from view, the sensitive devices are an acknowledgment that this seemingly serene plateau could blow so hard it would make the 1980 Mount St. Helens explosion look like a sneeze.

This summer, Yellowstone was added to the nation's handful of official volcano observatories. The others, smaller but far better known, are in Hawaii, Alaska, the Cascades and California's Long Valley. The Yellowstone observatory consists of a string of 28 electronic detection stations scattered through the park. Related plans call for at least 100 more monitoring sites.

Confirmation that the volcano was active was one of the most important factors in getting a new observatory established here.

What took so long for science to put its ear to the ground? For one thing, scientists couldn't decide whether the volcano was an infant or in its death throes. For another, it's just too big. The caldera stretches more than 30 miles from the north rim, a few miles from the Yellowstone River's Upper and Lower Falls, to the southern edge.

The last huge eruption was 640,000 years ago. Since then, a series of smaller ones have filled in the caldera "like tubes of toothpaste squeezing out all over the place," Smith said.

The Earth has always shaken periodically around Yellowstone. But without the proper monitoring equipment, no one knew how often it happened or why. Smith, who has been investigating here for more than 30 years, set up seismometers and found earthquakes by the hundreds. But the earthquakes — 15,000 between 1973 and 1998 — didn't fit conventional patterns.

Smith started thinking about the quakes in combination with Yellowstone's famously unstable plumbing. Was it possible that the quakes and the geysers were products of underground magma flows?

Smith and Robert Christiansen of the U.S. Geological Survey found evidence that a huge plume of magma rose from deep within the Earth and bore through the continental plate. As the plate moved southwest, the "hot spot" left a series of what Smith terms "ancient Yellowstones" across southern Idaho from Oregon to Montana.

In the mid-1970s, while surveying an old benchmark put into place when the first roads were cut through Yellowstone in 1923, Smith found that the ground had risen three feet in five decades.

There could be only one explanation. The volcano was bulging upward.

The caldera rose an inch a year until 1985. Then a swarm of earthquakes occurred nearby. By 1987, measurements showed that the caldera was falling an inch a year. In 1995, it started rising again. Now the caldera is bulging again, toward the southwest.

The scientists' work has yet to answer the most important question: Will the volcano blow its top again?

Christiansen doubts the likelihood of another cataclysmic eruption any time soon, but he doesn't rule out something smaller. A blowout on the scale of Mount St. Helens is conceivable, he said, adding: "We need to be prepared."

Daniel Masias (Muh Sigh Uh)

Published on July 28, 2000
Gazette Telegraph, Colorado Springs, Colorado

ATTENTION · Millennium 3000 DISCOVERY

THE MOST ASTONISHING WORLD WIDE (EGYPTIAN-GREEK-ASIAN-BIBLICAL) ARCHAEOLOGICAL DISCOVERY IN THE HISTORY OF MANKIND · HAS BEEN UN-EARTHED BY DANIEL. MASIAS, AN INDEPENDENT EGYPTIAN-GREEK RESEARCHER, FROM GREEN MOUNTAIN FALLS, COLORADO.

WHAT IS THIS DISCOVERY ABOUT AND HOW WAS IT UNCOVERED?

1. I looked at Mars (known to the ancients as the god of war and destruction) and found that 4 giant volcanoes there form a triangle.
2. From my research I knew that a number of ancient texts from the Egyptian Book of the Dead to the Bible, all mention geologic events such as lakes of fire, sulphur, smoking rumbling mountains and so forth. These geologic events are all associated with God.
3. My Egyptian studies also revealed that many tombs including the Rosetta stone were composed and built of granite basalt, and diorite - which are earth materials that are all volcanic in origin.
4. From the moon landings I was also aware that volcanic hot spots were discovered on the moon. The 3 hot spots on the volcanic moon form a triangle like the triangle on volcanic Mars.
5. From my Bible studies I knew that Jesus was a Jew and the religious symbol of the Jews is a geometric configuration, of 2 intersecting triangles. The Star of David also has 6 additional triangles, that travel in a clockwise 360° at the tips of the star. Because of the holocaust, I reasoned that the yellow star must have another meaning.
6. I also took into account the fact that the 15th century scientist Descartes proclaimed that God's existence could be demonstrated in geometrical fashion. A triangle is a geometric symbol, the Star of David
7. Finally in addition to other clues not mentioned here. I added the universal knowledge about the Bermuda triangle. In this triangle, ships, planes and people mysteriously disappear and death is final
8. Love triangle where 1 or 2 people die (history repeating itself).

WHAT DID I DO WITH THIS KNOWLEDGE?

I had an idea that I asked myself. What would happen if I took an Earth globe and made a red mark at all of the Earth's major volcanic-earthquake and temple sites around the world. When I did this I was astonished to see that 6 triangles formed all over the Earth, just like the triangles on Mars and the moon. When the Bermuda triangle is added there are 7 triangles all over the Earth. Mankind's entire existence depends on 7 continents as well as the 7 days or 365 days of the year orbiting the sun. The #7 is also associated with all sorts of religious themes and is considered the ultimate lucky number. The #7 is actually a 2 sided triangle with one side missing. Mankind's brain and soul which distinguishes him from animals, is manifested in the skull. The skull has 7 openings - 2 ears - 2 eyes - 2 nostrils - 1 mouth. Man's brain makes war possible, relating to the 7 triangles.

WHAT IS THIS ARCHAEOLOGICAL DISCOVERY ABOUT?

I have discovered that there are 6 more Bermuda Triangle images that exist all around the earth. On the boundaries of these 6 ancient triangle images, mankind's terrible wars, genocides, plagues, disasters and natural calamities have been taking place for many thousands of years. Just as death takes place in the Bermuda Triangle, death also occurs on these 6 mysterious triangles.

WHERE ARE THESE MYSTERIOUS TRIANGLES LOCATED ON EARTH?

(Note: take a red magic marker and a globe of the earth and draw a red dot at all of the world's major volcanic, earthquake and temple sites around the world) The following locations are those sites.
Triangle #1: Giza Egypt to Aztec Mexico City to Inca Peru.
Triangle #2: Kamchatka peninsula to Tambora Indonesia to Easter Island.
Triangle #3: Italy to Afghanistan to the Aleutian Islands.
Triangle #4: Equatorial Guinea Africa to Tristan Da Cuna to the Maurillus Islands near Madagascar
Triangle #5: Colorado to Washington State to Baja California (La Paz) Colorado: Volcanoes, Hot Springs, San Creek Massacre, Big Thompson Flood, 737 Crash, Suicide, Hail, Albuquerque Lows.
   Wyoming - Montana: Yellowstone Volcanic Activity, Cheyenne, Sioux Indian Wars, Custer's 7th Cavalry Battle. Idaho: Craters of the Moon National Monument, Volcanoes, Hot Springs. Washington: Mt. St. Helens Eruption 5/18/80, Mt. Rainier, Ted Bundy Killer, Hanford Nuclear Pollution. Oregon: Volcanoes, Mt. Hood, 3 Sisters Volcanoes, Green River Killer. California: Volcanoes, Mt. Shasta, Mt. Lassen, Mammoth Lakes, San Francisco Great Earthquake, Mass Killers, Plane Crashes, Great Fires, Rain Baja California: Volcanoes, Mysterious Disappearance of Ancient Native Peoples. N.W. Mexico: Volcanoes, Pancho Villa Revolution, New Mexico & Arizona: Volcanoes, Wyatt Earp, Tombstone, Boot Hill, Geronimo, Cochise, Apache Wars, Billy the Kid, Lincoln County Wars, the Atomic Bomb, the Roswell UFO Crash, the Los Alamos fire. Colorado: Great Sand Dunes, Ancient Egypt
Triangle #6: This triangle #6: This triangle is formed in western Europe by the intersection of Triangles #1 and 3 which equals 4. Mars is the 4th planet and the god of war. The 6th triangle correctly identifies man's most horrible holocaust and terrible wars, genocides and plagues.
Triangle #7: The Bermuda Triangle, off the coast of FLorida. *Source: Groliers Encyclopedia, the New Book of Knowledge, 1970 Vol. 19 P385 (NOTE: History repeats itself because of these triangles)*

WHERE DID THIS CURSE COME FROM?

In Genesis God put a curse on mankind because of the misbehavior of Adam and Eve. The first murder happened when Cain killed Abel. All around the world now, brother is killing brother. This curse is associated with the major volcanic regions of the earth and man's temple sites. God's curse is in the physical shape of a rare volcanic material that emerges from only 3 places on earth. This material is called hexagonal (6) prismatic columnar volcanic basalt. Large amounts of this rare material is located at Devil's tower in Wyoming, Nan Matol in Micronesia and the Giant's causeway in Ireland. Each location is the scene of past great disasters, from World War II Pacific battles to European wars of unimaginable carnage; to the North America disasters of the American Civil War, Indian genocide and continuing tornado disasters, and floods. Each location has a 666 element of disaster, when we draw 6 triangles onto the 6 sided hexagon.

THE GREATEST GLOBAL ARCHAEOLOGICAL DISCOVERY IN THE HISTORY OF MANKIND

Two years ago on January 7, 1998 my research was published in this newspaper and it predicted that global disasters and heinous crimes would occur on these triangle lines-all over the earth. The 2½ year public test of this discovery has turned out to be 100% correct!

80

THE WHITE HOUSE CONTACTED ME AND CALLED MY RESEARCH QUOTE "A GLOBAL ARCHAEOLOGICAL DISCOVERY." THE BRITISH GOVERNMENT CONTACTED ME AND HAS TAKEN NOTE OF THIS MATTER. SENATOR WAYNE ALLARD WROTE TO ME AND EXPRESSED HIS INTEREST. BECAUSE OF THIS AMAZING DISCOVERY I AM PLANNING TO PRODUCE A FILM ABOUT THIS RESEARCH, IF WE CAN RAISE ENOUGH FUNDS TO PRODUCE IT. PLEASE DONATE TO THIS WORTHY CAUSE ON BEHALF OF MANKIND.

THE FOLLOWING DISASTERS ALL HAPPENED ON THE #5 TRIANGLE WHICH STARTS IN VOLCANIC COLORADO AND TRAVELS TO WASHINGTON STATE. AND BAJA CALIFORNIA AND THEN GOES BACK TO COLORADO. THERE WERE MANY MORE DISASTERS NOT MENTIONED HERE

1. Columbine high school massacre in Littleton Colorado	April 20, 1999 GT
2. Vail lodge fire, greatest eco-terrorism in US history	Oct. 23, 1998 GT
3. La Junta Colorado flooding and Manitou flooding	June 7, 1999 GT
4. 7 killed on I-25 near Briargate Colorado Springs	Feb. 7, 1999 GT
5. 7 killed in Denver by a serial killer	Nov. 18, 1999 DP
6. 3 killed on Texas church bus near Canon City Colorado	Dec. 22, 1999 GT
7. 3 killed in van near La Junta Colorado 7 injured	Jan. 23, 2000 GT
8. Heinous killing of Matthew Shepard in Wyoming	Oct. 13, 1998 GT
9. Air Force plane crash in Idaho	Jan. 21, 2000 GT
10. Aryan Nations conflict with US in Idaho and Montana	Feb. 20, 2000 DP
11. Salt Lake City tornado disaster in Utah	Aug. 12, 1999 GT
12. Mormon Church and TV station shootings in Salt Lake	Jan. 5, 1999 GT
13. 6 killed in refinery explosion in Washington State	Nov. 27, 1998 GT
14. 10 million dollar WTO riots in Seattle Washington	Dec. 5, 1999 GT
15. 1999 disastrous summer forest fires in California, Idaho, Oregon and Nevada	July 7, 1999 GT
16. 7 dead in Marine copter crash near San Diego	Dec. 10, 1999 GT
17. Jewish's children's school shooting in Los Angeles	Aug. 11, 1999 GT
18. Alaska Airlines MD 80 plane crash in the Pacific (LA)	Feb. 1, 2000 GT
19. 6 killed in Jean Nevada near Las Vegas in a van crash	Mar. 9, 2000 GT
20. 21 killed in Baja California by gunmen	Sept. 18, 1999 GT
21. 10 killed in church bus on New Mexico road	May 3, 1999 GT
22. 9 killed in a van accident on road in New Mexico	Dec. 5, 1999 GT
23. State wide power outage in New Mexico	March 19, 2000 GT
24. 19 killed in marine plane crash near Tucson, Arizona	April 8, 2000 GT
25. Los Alamos New Mexico Fire	May 7, 2000GT

THE FOLLOWING DISASTERS OCCURRED ON THE #1 TRIANGLE FROM AZTEC MEXICO CITY TO GIZA EGYPT TO INCA PERU AND THEN BACK TO AZTEC MEXICO CITY AND THE EUROPEAN TRIANGLE.

1. The death of John F. Kennedy Jr. and his family in his plane	July 17, 1999 GT
2. The Swiss airplane crash off of Nova Scotia	Sept. 3, 1998 GT
3. The Egypt airplane crash out of New York	Nov. 1, 1999 GT
4. The severe 1999 northeast U.S. drought	July 24, 1999 GT
5. Mass graves found at Ciudad Juarez Mexico-Texas	Nov. 30, 1999 GT
6. 10 clammers lost at sea near New Jersey	Jan. 21, 1999 GT
7. Devastating earthquakes in Turkey and Greece	Aug. 18, 1999 GT
8. 6 firefighters killed in Boston fire	Dec. 5, 1999 GT
9. 11 Texas Aggies killed in collapse of pyramid logs	Nov. 19, 1999 GT
10. 60 dead in dance hall fire in Sweden	Oct. 10, 1998 GT
11. 33 dead in Oslo Norway train crash	Jan. 5, 2000 GT
12. 1000's dead in South African flooding	Feb. 12, 2000 GT
13. 650 killed by mass murders in Uganda	March 21, 2000 GT
14. U.S. led Nato air war in Yugoslavia	Oct. 7, 1998 GT

DISASTERS ON TRIANGLE #2 (SEE LOCATIONS ABOVE)

1. 7 dead in Hawaiian Xerox plant, shot by gunman	Nov. 3, 1999 GT
2. 7 dead in Hawaiian landslides	May 11, 1999 GT
3. Incredible devastation of hurricane Mitch in Central America, Triangle #1	Nov. 3, 1998 GT
4. 2 American embassies blown up in Africa	Aug. 8, 1998 GT

Mankind must learn how to live in peace and harmony with others and have tolerance for others. We must all be brothers and sisters.

THE IMPORTANCE OF THIS DISCOVERY

This research is as important as the works of Copernicus, Galileo, Kepler, Descartes, Newton and Einstein. It is worthy of a Pulitzer Prize and The Nobel Peace Prize. The 2 1/2 year public test of this global archaeological discovery turned out to be 100% correct.

DANIEL MASIAS, PRONOUNCED MUH-SIGH-UH GREEK PRONUNCIATION, MESSIAS

THE FILM

WE ARE PLANNING TO PRODUCE A FILM ABOUT THIS DISCOVERY. PLEASE HELP US TO HELP MANKIND TO UNDERSTAND WHAT IS HAPPENING HERE ON THIS PLANET, BY DONATING $100-500-1000-? TO MAKE THIS PROJECT HAPPEN. YOUR DONATION NO MATTER HOW SMALL WILL BE ACKNOWLEDGED IN THE CREDITS OF THE FILM ALONG WITH A PERSONAL LETTER OF THANKS FROM ME WHEN THE FILM IS COMPLETED. AT THE COMPLETION OF FILMING, THE MOVIE WILL BE SHOWN AT THE TELLURIDE FILM FESTIVAL AND THE SUNDANCE FILM FESTIVAL. PLEASE HELP US BY DONATING TO THIS VERY WORTHY CAUSE ON BEHALF OF MANKIND'S FUTURE. IF YOU KNOW OF AN UNUSUAL PLACE TO FILM, OR YOU HAVE A BEAUTIFUL GARDEN OR STONE WALL OR HAVE LARGE ROCK FORMATIONS, CONTACT US FOR A POSSIBLE FILMING SITE.WE NEED ACTORS, ACTRESSES, SINGERS, MUSICIANS AND GENERAL FILM PEOPLE FOR THIS FILM PRODUCTION. WE NEED VOLUNTEERS OF ANY TYPE TO ASSIST IN THIS FILM. IF YOU WANT TO WORK PRO-BONO IN ANY CAPACITY, PLEASE FEEL FREE TO CONTACT DANIEL MASIAS AT 884-9468 OR SEND YOUR RESUME TO DANIEL MASIAS S P.O. BOX 14, Green Mountain Falls, Colorado 80819.

THE MESSIAH ASSOCIATED WITH THE MOON

from the earth to be lined up. According to the first century historian Josephus, these seven heavenly bodies represent the menorah in the Temple (antiquities 3, 11:7). Not only does the moon represent the Messiah in the Talmudic tradition, but the center lampstand of the menorah does, too. The three candle branches on either side of the center stem represent man grafted into the Messiah. Is our Heavenly Father recreating a symbolic menorah in the skies on this date? Consider also that on this date Solomon started to build the first Temple and Ezra began building the second. Does this signify the imminent building of the third Temple mentioned in II Thessalonians 2:4?

Is it any wonder we should be

(Revelation 12:1). Remember, the moon is very significant to the Jews. It is the basis of their calendar system and symbolizes the Messiah.

Now comes the best part. On May 7, 2000 or Iyar 2576 on the Hebrew calendar, Venus, Mercury, the Sun, Jupiter, Saturn, Mars, and the Moon will all appear

witnessing similar, perhaps even more dramatic, signs in the heavens if we are indeed on the threshold of the return of our Lord? Jesus Himself told us to expect such things (Matthew 24:29).

At the same time, the Bible warns against attempts to set dates for the return of our beloved Savior. We should scrupulously avoid such unscriptural practices. Yet, the Word of God commands us to recognize the general "signs of the times" which will herald His coming. Some spectacular signs are occurring right now. Look up, our redemption draweth nigh! (Luke 21:11, 25).

or rain and produces pellets or grains of ice. These are commonly referred to in North America as *sleet* but, among the British, "sleet" refers to a mixture of snow and rain.

Hail, another form of precipitation, consists of large pellets or spheres of ice. The formation of hail will be explained in our discussion of the thunderstorm.

FIGURE 6.13 These individual snow crystals, greatly magnified, were selected for their variety and beauty. They were photographed by W. A. Bentley, a Vermont farmer who devoted his life to snowflake photography. (National Weather Service.)

Daniel Masias (Muh Sigh Uh)

Daniel Masias (Muh Sigh Uh)

Devil's Tower Wyoming

OFFICE OF THE VICE PRESIDENT
WASHINGTON

December 10, 1998

Daniel A. Masias
P.O. box 14
Gr. Mtn. Falls, CO 80819

Dear Mr. Masias:

I want to thank you for taking the time to send me information regarding your global archaeological discovery. I will keep your material on file as part of a growing library of reference materials.

I am encouraged by the many Americans who have shared with me information and resources regarding their scientific researches. It is good to know the Clinton-Gore Administration has your support in this endeavor.

Sincerely,

Tipper Gore

Daniel Masias (Muh Sigh Uh)

British Embassy
Washington

3100 Massachusetts Ave. N.W.
Washington, D.C 20008-3600

Telephone: (202) 588-6593
Fax: (202) 588-7889
E-mail: david.arkley@
washington.mail.fco.gov.uk
Web-site: http://www.britain-info.org

5 October, 1998

Daniel A. Masias
P.O. Box 14
Green Mountain Falls, CO 80819

Dear Mr. Masias

Thank you for your letter to the Ambassador of 27 September. I have been asked to reply.

The points you raised in your letter have been noted. Thank you for taking the time to bring this to our attention.

Yours etc.

David B. Arkley
Third Secretary Press and Public Affairs

WAYNE ALLARD
COLORADO

Hart Senate Office Building, Suite 513
Phone (202) 224-5941
Fax (202) 224-6471

BANKING, H
URBAN

ENVIRON
PUBLIC

INTEL
CIPA

United States Senate
WASHINGTON, DC 20510-0606

December 1 , 1998

Mr. Daniel A. Masias
Post Office Box 14
Green Mountain Falls, Colorado 80819

Dear Daniel:

Thank you for forwarding the results of your global
archaeological research. I appreciate you taking the time to
share this information with me.

As you may know I am a veterinarian. Because of my scientific
background I am always interested in the forces that shape the
earth and geological events. You have presented some intriguing
thoughts in your letter and I wish you well with your continued
research.

Sincerely,

Wayne Allard
United States Senator

WA:rmb

Daniel Masias (Muh Sigh Uh)

Earth, The Water Planet And The Ancient Classic Style

One of the keys to understanding mans ancient history is that water, oceans, volcanoes, earthquake zones, ancient pyramids and temple sites all over the earth are some of the critical elements to understanding our very mysterious past history. When we look at the earth, we are not surprised to learn that the earth is drenched in water and of the nine planets of our solar system, we live on the only planet that is literally drenched in water and water vapor in it's atmosphere. The earth is 70% water based and 30% land mass. The human body is also 70% water based and 30% bone calcium, which is the same percentages as the earth. The white bone calcium is similar to the sea shell calcium, that sea creatures such as shells are made of humans are encouraged to take calcium supplements before old age, to stem the effects of osteoporosis which is a loss of bone calcium. Every human being on the planet is biologically linked to the oceans and the seas of the earth. We humans sweat a salty liquid called perspiration and our blood is salty tasting.

Humans are conceived when a small spermatozoa, which looks remarkably like a "tad pole which is an amphibian," swims by its tail motion to the female egg and fertilizes it, by penetrating the woman's egg. Research has shown that as the human fetus develops in the womb, there is a remarkable stage at which the human fetus possesses amphibian like, slightly webbed hands and feet. This confirms our ancient ancestors claims that mankind is intimately connected to the earths oceans and waters, as well as The Masters of The Waters.

Like lemmings that flock to the oceans and seas, humans all over the earth, and for countless centuries have practiced this same ritual. The need to be near the oceans and sea's, in order to connect back to our original beginnings. This is a deep need of humans all over the world, that is buried in the remotest realms of our psyche that beckons us to return to our creation cradle, which is the waters of the earth. Confirming our water drenched world, the earths skies are blue and over cast gray; which are the various colors of the earth's ocean waters. The sky is blue because the white light of the sun strikes water molecules, gases and the dust in the atmosphere. Because of the large amount of water vapor in the atmosphere, the white light is broken up into the colors of the rainbow; and the blue and violet is more scattered than the other colors. Because of this effect, the skies of the earth are blue, just as its waters are blue. When we look at the rainbow, we remember the same colors, and the shape of this water bow, is the same shape as a sea shell.

Humans have a ritualized need to gravitate to the oceans and sea's of the earth, which is the source of our very beginnings, and is our very sustenance. When sea life is in trouble or when whales are beaching themselves, there is a great out pouring of human kindness and help to save their lives and return them to the ocean. When we look at the earth and the great amounts of water on it's surface, we are amazed to discover that other planetary bodies of our solar system also have oceans and sea's, that are frozen. The Voyager spacecraft in 1979 discovered large amounts of frozen water on Saturn's moon Titan. The Voyager spacecrafts also discovered water on Jupiter's moon Europa, and the first active vulcanism ever discovered outside of the earth, was found at Jupiter's moon Io.

In a June 1993 article in Parade magazine, astronomer Carl Sagan stated that water is a very abundant and common element in the universe. As we are all aware, life would be impossible on the earth without water.

The symbols of water are everywhere in the ancient history of mankind. From the building materials for their ancient temples and pyramids, such as marble and limestone; which have their origins in the oceans to the architectural ocean and sea depictions of the Greco-Roman building styles. Everywhere there exists the themes of the earth's waters, fish, sea shells, ocean scrolls as well as other symbols, throughout the ancient world. The symbols of the waters were also carried over to their sculpture, art, furniture, literature, and even the way and manner in which they curled their hair locks.

A number of man's ancient cultures and civilizations all inform us that superior beings associated with fish, amphibians, oceans and waters of the earth, and the star system Sirius came to earth and taught civilization to our ancient ancestors. Our ancient ancestors also tell us that in fact mankind and all of the life on this planet, are the direct product of these mysterious beings from Sirius.

The legends of ancient man from all over the world inform us that the primary numbers 2,3,5,7, and 9 were all sacred and powerful numbers. From the ancient Babylonians to the Greeks and the Romans, these numbers were all considered important, magical, powerful and sacred. The ancient Egyptians as well as other civilizations built their temples and pyramids such as the three great pyramids at Giza, to face east to the star Sirius. By facing east, I mean they placed the front entrances and positions of their temples to face the brightest star in the eastern sky. This star was Sirius, a magnitude 1 star that appears like a headlight in the eastern sky from July to September. Sep was one of the ancient Egyptian Gods, and these three summer months are known world wide as the dog days of summer, because they are known as hot muggy days and are associated with the helical rising of Sirius in our eastern skies. Of the seven wonders of the ancient world, only the three great pyramids at Giza are still standing in their ageless magnificence. Their very existence is telling mankind an incredible story, but man has not wanted to listen to that story. The mysterious construction of the great Egyptian pyramids and the entire religious worship of it's pantheon of Gods, was all for the expressed purpose of worshipping the star Sirius. During the ancient Egyptian dynasties, the black Kushite rulers of Nubia, also built pyramids and tombs in worship of the superior beings of Sirius. The Kushite pharaohs of Nubia ruled ancient Egypt during the end of the last of the 30 dynasties, of ancient Egypt. The modern Dogon people of Mali Africa are the descendants of the Kushite. Before and after the construction of the three great pyramids at Giza, our ancient ancestors recorded in cuneiform, clay tablets, papyrus, stone glyphs, and other ancient artifacts that superior beings from Sirius were called the Nommos and they were also known as the superior being called Oannes. Other civilizations around the world from indigenous Indians of the Americas such as the Anasazi, Aztec, and Incas to the Island peoples of the world all spoke of these same superior beings. They came from the skies in shiny luminous disks. They had different names for these beings, but they were the same humanoids from the stars. To our ancient ancestors these superior beings were associated with fish, amphibians, oceans, sea's and the waters of the land. They are called The Masters Of The Waters, The Instructors, The Monitors. Our ancient ancestors inform us that Oannes as well as his fellow beings, came from Sirius and that they live on a giant Jupiter sized planet, that is covered by water like the earth. These beings like humans of the earth are in fact water creatures, but they have mastered the secrets of the universe and they have conquered space and time unlike the human race. Ancient written artifacts and modern traditional religious worship of the Dogon people, inform us that the Nommos or Oannes and his fellow beings brought humans and the flora and the fauna of Sirius to the earth. Mankind was instructed in all the aspects of civilization, and how to live in peace among one another. Oannes taught men how to construct homes and temples and how to till the fields of the earth. Men were taught laws, geometry, and astronomy as well as given instructions in how to be civilized. Many of our ancient ancestors said these events had taken place over a period of five civilizations on the earth. Currently, according to our ancient ancestors mankind is living in the fifth civilization. As the monitors of mankind our ancestors inform us that these beings have been watching over the peoples of the earth for eons of time. The beings who operate the flying disks in the skies over the earth, are the humanoid creatures called Bes. They are on a different order than Oannes, and there exists a number of ruling entities that govern their realms.

Mankinds ancient ancestors left written artifacts that tell us that the road to the real history of man, lies with the superior beings of Sirius and the oceans and sea's of the earth and Sirius as well as the superior beings who reside there. All of our ancestors from the Babylonians, Summerians, Assyrians, Minoans, and the Egyptians to the Greeks and the Romans all worshipped these Gods from Sirius. In the ancient world of Egypt, Babylon, Greece and elsewhere, only the secret religious initiates knew the entire true history of the beings from Sirius. The average ancient citizen knew of the Gods, but did not know the detailed story of who they were or where they came from. All of this information was kept secret by the Pharaoh and his specially chosen secret priests. This information was kept as highly guarded secrets in their temples, in much the same manner that modern governments today keep secrets from it's citizens. The same was true with other peoples and civilizations. Only a handful of secret religious initiates had in their possession the full knowledge of the beings from Sirius. Just as the ancient priests were aware that mankind was brought from Sirius to the earth, which explains why modern anthropologists cannot find the missing link we also learn that the whole human concept of keeping secrets from the general population, has followed mankind down from the time of the Pharaoh's.

The ancient Egyptians and other peoples greatly influenced the Greeks and the Romans, in their worship of these superior beings. The Rosetta stone with it's Egyptian hieroglyphics and Greek writings below, are testimony to that fact. From the classic Greco-Roman architecture of scrolled columns, rosettes, volutes, scrolled capitals on the top of the sacred temple columns and other ocean themes on their sacred temple buildings the Greeks as well as the Romans worshipped the superior beings from Siryius, in an allegorical and arcane manner.

To understand what is meant by the classic style of architecture, we must have a geologic definition of an ocean scroll. The modern and ancient geological term for ocean scrolls are long curving parallel ridges created by the movement of water. Scrolls are defined by the Websters's New Lexicon dictionary as quote: A spiral design which resembles the end of a loosely wound scroll they appear in Greek and Roman architecture. Throughout the ancient world the classic architecture style Roman civilization personified the oceans, sea's and waters of the earth as the identifying mark of the superior beings from the star system Sirius. The classic style of man's ancient past has for thousands of years inspired great masterpieces in sculpture, architecture, painting, music, and literature. We thus learn that The Masters Of The Waters from Sirius were the source of the ancient classic styles of the Greco-Roman eras. The most important sacred temples that were built to the Gods such as Athena, Zeus and so on, as well as their government and private villas were all constructed in the classic style. That classic style envolved the three orders of stone building columns that were referred to as the Doric, the Ionic, and the Corinthian styles. The three orders were associated with the spiral scroll volutes that are on the top or capitals, of these stone building columns. These were the columns that held up the stone roofs of the sacred temples that were expressly associated with the earths oceans, sea's seashells, amphibians, rosettes, fish and the salt of the earth. In short just as the ancient Babylonians, Summerians, and Egyptians built pyramids and temples to The Masters Of The Waters from Sirius; so too did the ancient Greeks and the Romans.

Other ancient peoples and civilizations of the world also worshipped the superior beings from Sirius. The ancient Aztec's inform us that the stars determined the fate of the world, referring to Sirius. The Pleiades a cluster of seven stars was very important to their world. Just as the Pleiades was very important to the Aztec's, it was also very important to the Greeks as well. We learn that the Pleiades is a northern constellation with seven bright stars in the Taurus constellation. In Greek mythology these seven daughters of Atlas were associated with Zeus. Maia, the eldest of the seven daughters was the wife of Zeus, and together they were the father and mother of Hermes the messenger of the Gods.

The Ancient Dogon People

We now turn our attention to the Dogon people of present day Mali Africa. They are the descendants of the Kushite rulers of Nubia or Kush. They revealed a short 43 years ago in the 1940's, a series of astonishing revelation of their religious worship. The Kush were ruled by the Egyptians and they adopted the religious worship of Sirius, that the ancient Egyptians have been practicing for millennia. The Kushites kept the religious worship of Sirius alive for thousands of years through their descendants, the Dogon people. Like the ancient secret religious initiates of Summer, Babylon, Egypt, Greece and else where, the Dogon secret religious initiates also kept the secrets of Sirius to themselves. It was not until the middle of the 20th century that the Dogon people revealed the full implications of their secret religious worship of Sirius. In the 1940's after many years of anthropological work among the Dogon people, two French anthropologists were told about the ancient secrets of their religion. The ensuing information turned out to be astonishing. The religious initiates revealed that their religious worship of Sirius has a history that goes back in time thousands of years. They revealed that visitors from the star Sirius came to the earth eons ago, and made contact with their ancestors. The Dogon people called these beings the Nommos. They are described as part amphibian beings who were sent to the earth for the benefit of mankind. The ancient Dogon people, and their knowledge of Sirius inform us that the Nommos brought man as well as all the animals and plants of Sirius to the earth. The Nommos revealed to the Dogon, astronomical information about Sirius that was not able to be confirmed by astronomers until 1970. Before that time, scientists were not able to confirm certain scientific facts about Sirius. Astonishingly, the facts that the neolithic Dogon people revealed, turned out to be absolutely correct. How could a primitive African people who live in caves and straw huts and have absolutely no modern technology or scientific skills, much less telescopes possess correct scientific information about Sirius? A star system that is 9 lights years from the earth!

The Dogon people were also told about the rings of Saturn and the four major moons of Jupiter, as well as other astronomical facts that are correct. All of these scientific facts were given to the Dogon people by the beings from Sirius. The Dogon people informed us that these beings are The Masters Of The Waters, The Instructors and The Monitors of all mankind.

We therefore learn that Oannes who the Babylonian priest Berossus wrote about as being from Sirius was the father of mankind; was therefore one and the same as Oannes. Other peoples and civilizations from around the world also spoke of these superior beings from the stars, and we know them as the beings from Sirius.

In the final analysis, mankind's ancient ancestors reveal to us through the three great pyramids, marble and limestone temples, edifices, stone glyphs, clay tablets, tomb hieroglyphics, papyrus, cylinder seals and millions of other ancient antiquities(as well as the ancient worship of Sirius, that the entire history of mankind is the result of the superior beings from the star system Sirius.) Just as the earth is drenched in water and is the water planet, mankind is biologically linked to the seas, fish, amphibians and the waters of the earth. We also learn from our ancient ancestors that the beings from Sirius also came from a giant water planet, and these beings are the mentors of mankind. They are The Masters Of The Waters, and they come from the stars.

Our Ancient Ancestors

Ancient peoples all over the earth inform us that mother earth was extremely important to their lives. Pagans in Europe and elsewhere, as well as ancient indigenous Indian peoples and Island peoples all over the earth all worshiped mother earth as the source of their well being and the very fabric that unified peoples all over the earth. Through their beliefs and ancient religious practices we learn that mother earth was a living breathing entity that protected all life on the planet and was bequeathed to mankind by the superior beings from Sirius.

Our ancient ancestors derived their very lives and physical existence from the bounty that mother earth provided to them. From the crops that they grew and harvested to the animals that they consumed as well as the trees, rocks, hay, and mortar from the earth mother earth was worshipped and revered as the provider of man's very sustenance.

Modern man's relationship is no different today because he too lives off the bounty of mother earth, though at a higher level of technology and sophistication. Every human being owes his very physical lives and existence to mother earth, because food crops are grown from the very dirt of the earth. The very dirt of the earth is the source of all of mankind's food sources, as well as his other basic needs for survival. Indeed the Earth is our modern sustenance, yet we have totally forgotten about how important she is to our lives and spirits.

Thus we learn that the earth and it's waters are the protectors of all life on the earth. To the ancient Greeks mother earth was known as Gaia and the Romans referred to her as Terra. The earth was therefore revered by ancient man and mankind lived in harmony with his surroundings.

The role of ancient women was equated to that of mother earth, thus they played a pivotal role in ancient times. Ancient women were also known to be important in other aspects and legends. Ancient Summerian clay tablets as well as other sources inform us that The Sons Of The Gods who came from the stars, took the beautiful women of the earth. The beautiful women who were taken by The Sons Of The Gods, then bore the children of their unions with these superior beings. These children were called the Demi Gods and we know them as Hercules, Jason, Atlas and others. These Demi Gods competed in physical games of skill and strength against the biggest and best of earths mortals.

These games later came to be known in Greece as the Olympic games. Today's modern Olympics and worldwide sports activities, are the results of the competitive spirit between the ancient Demi Gods and the best athletes of the earth.

Because the women of the earth bore the children of The Sons Of The Gods, women for thousands of years have always consulted the stars and astrology to glean their futures in the positions of the planets and the stars. Some women have subconscious race memories that extend back in time eons ago. They remember a time deep in the recesses of their minds, when The Sons Of The Gods came to the earth, and the memory of these superior beings has been passed on from generation to generation. It is because of the memory of these beings from the stars, that, mankind today has the Zodiac, astrology and the horoscope. These ancient oracles intimately bind mankind to a time deep in our ancient past, when The Sons Of The Gods competed with ancient man in the proto Olympic games, and mingled with the beautiful women of the earth.

At a recent meeting in May 1992, of the American Association for the advance ment of Science; they reported that in a number of studies and surveys that more than one third of American adults believe that astrology has some scientific merit. Many of these Americans are women who believe in the reading of their horoscopes in the United States and other Western countries.

Today this ancient concept and belief that certain people of the earth are some how "higher" than other people is very much alive and well all over the world.

Because of this fact that certain people are perceived to be of a better birth right, this could well be a subconscious feeling in many people around the world which has led to many of mankind's problems concerning racism.

From almost the beginning of the twentieth century; and the advent of the motion picture industry; people such as actors and actresses who played the leading roles in the movies or plays have always been referred to as Stars. The reference to these physically beautiful men and women as Stars, originated with the subconscious race memory of The Sons Of The Gods. This human concept is in the same category as that of royal kings and queens. Therefore like the proto Olympic games of ancient Greece, and the modern Olympic games as well as mankind's world wide sports, The Sons Of The Gods still exert a great deal of influence on modern man.

The importance of earth women and queens to The Sons Of The Gods is therefore not lost to us. Like mother earth who watches all life on the earth, women in ancient times were very important to society, and they took on the responsibilities of mother earth in protecting and caring for their families. Because our ancient ancestors knew that there were superior beings from the skies, and they were the protectors of mankind they connected mother earth with these superior beings from the stars and they therefore worshipped them as their protectors.

The Rosetta Stone And The Trismegistic Literature

There exists ancient written records as well as world wide physical signs that proves the existence of the mysterious beings from Sirius and their influence on mankind's history and the earth. These worldwide signs are the representations of the oceans, sea's, fish, pyramids, the dog, cow, bull, as well as other physical signs of the world, from ancient times as well as today. From the ancient Minoans and their worship and veneration of the cow and the bull, to other ancient civilizations around the world we learn that the bull as well as other animals were sacred and they commanded great respect and veneration from our ancient ancestors.

The ancient Egyptians provide one of the most important sources of information that reveals the existence of the beings from Sirius. One of these sources is called the divine eye of Ra the Egyptian sun God. This refers to a single large oval eye that is portrayed on many of the tomb paintings of the Pharaoh, Osiris and other Egyptian Gods.

The Egyptian Pantheon of Gods that are portrayed on the tomb walls almost always show a single oval eye. Ultimately this single eye was referred to as the Divine Eye of Ra the Sun God. Later I will refer back to this single oval eye in the context of the American one dollar bill, and the Nostradamus prophesy.

Another very important source of information comes from the ancient Babylonians and their secret initiate priest named Berossus. Only fragments of his history of Mesopotamia have survived the ravages of time. These fragments were in turn recorded by later Greek historians, such as Apollodorus. Much of what Berossus wrote was deliberately destroyed at the great library of Alexandria. The fragments that did survive however, and which have been hidden away in large libraries, reveal a totally different story about the history of mankind. The fragments of Berossus have in effect been hidden away for over 2,000 years because there was a lack of scholars and publishers who, were interested in the writings of this Babylonian priest. Those scholars that have researched this ancient priest have been prejudiced in their analysis of his written works, in favor of the more traditional views of man's history and all that it entails. They have performed a great disservice to mankind. There is no doubt that the history of Mesopotamia as told by Berossus, goes against the traditional thoughts of Western man. There in lies the biggest reason that Berossus has been shunned, castigated, and ignored by traditional historians and scholars. The underlying current is that the history that Berossus and others wrote, poses a great threat to mankind's traditional historical beliefs. The same arguments and statements can also be said about the ancient Egyptians and their worship of Sirius because very few people outside of historians and scholars know that the Egyptians worshipped superior beings from the star system Sirius. Even fewer people are aware of the fact that all the priests of the Mediterranean as well as the Levant and northern Africa, all worshiped the Gods through ancient secret religious ceremonies, that were kept hidden from the general population The implications of the ancient worship of Sirius has astonishing meaning for all of modern mankind.

A third great source of information comes to us from an ancient source called The Trismegistic Literature or (Hermes) and The Virgin Of The World. The Hermetic (Hermes) Trismegistic Literature is very important to understanding who the beings from Sirius are and the power and knowledge that they command. Like the fragments of Berossus that have been sequestered away, The Trismegistic Literature and The Virgin Of The World have been relegated to the same status. All of the information contained in the three sources of ancient knowledge, talk about the origin of mankind on the earth; and it presents us with information on who and what the beings from Sirius are and what will happen in the future as spoken of by it's prophesy.

When one looks in the 20 volume Oxford English dictionary, which is the officially recognized source of the English language, and is the source of the origin of English words and names we find that neither the Babylonian priest Berossus or the being known as Oannes can be found in the massive Oxford dictionary source. This is despite the fact that these two ancient names are in fact known to modern historians and scholars. The Trismegistic Literature as well as Berossus and Oannes, cannot be found in the English dictionary either. The only mention of the ancient literature is made in association with Hermes, the Greek God of wisdom and the messenger of the Gods. This is because Hermes is part of ancient Greek mythology and as such is associated with Western Civilization, because of it's ties to democracy and the classic style however when it comes to the ancient civilizations of the Mespotamians, Egyptians, Babylonians and other peoples of the Middle East, the English dictionaries do not mention any of these extremely important figures of the Sirius religion.

Various English dictionary sources label The Trismegistic Literature as being of the occult, which to some people means the devil and evil. Nothing could be further from the truth, and I suspect that through the centuries efforts have been made to deliberately label it as such.

Trismegistic Literature concerns the hidden secrets of astronomy, space science, solar systems, planets, stars and the unknown secrets of the universe; that scientists have been searching for. This literature has to do with the secrets of the cosmos, and the eventual discovery of the secrets of the star system we know as Sirius.

We therefore learn that Berossus and Oannes, as the chronicler's and the sources of mankind's true history have been censored out of man's literature and libraries so to speak, in favor of the more traditional theories on the origin of man and the earth. Mankind has therefore been denied a differing view of history that only scholars and historians have been aware of. These scions of modern knowledge and enlightenment as well as their predecessors have all pushed this important knowledge under the rug, so to speak and out of the sight of people. The ancient literature of Berossus and the written records and words of Oannes have however survived to this day although they are incomplete and despite the fact that efforts were made to destroy them. From The Trismegistic Literature we learn that before Oannes returned to the stars, they left behind on the earth un findable and hidden instruments. In order to investigate the possibility that hidden and mysterious instruments were left on the earth and were disguised from the eyes of man, we must first investigate the great pyramids at Giza and the ancient religious worship of Sirius. In this context we look at the three aspects of Egyptian history, and one of which was only deciphered in the early 19th century. This ancient artifact is the Rosetta Stone which was found in 1799 by a French engineer of Napoleon's army. It was partially buried in the muddy waters of the Nile River at Rosetta Egypt. It was quite appropriate that the stone was found in the waters of the mighty river, because this is exactly what mankind and the earth is all about. The Frenchman found the 2 foot by 3 foot black volcanic basalt stone tablet near Alexandria Egypt, site of the legendary great library of Alexandria which was a location in antiquity of great learning and knowledge. The library was later sacked and burned by savage hordes.

The Rosetta stone consists of a hard black basalt volcanic material which is of igenious origin or coming from a volcanic eruption. On the stone there are three distinct languages that are inscribed on the hard stone surface. the first inscription is the ancient Egyptian hieroglyphics language, while the second is the demotic or the more popular language of the ancient Egyptian people of 200 B.C.. On the bottom of the black stone is written the same dissertation in the Greek language. The volcanic stones three distinct inscriptions describe a decree that was written by the priest of Memphis in 196 B.C.. Ptolemy Epiphanes is honored for his various contributions to the priesthood as well as for his concerns for the average Egyptian citizen.

Rosetta Egypt and the Rosetta stone are the first two parts of our triangle or pyramid image, while rosettes are the third part of the pyramid that allows us to understand our mystery. Rosettes in ancient times were copied from ocean sea shells and the intricate designs have been reproduced by artists ever since. The rosettes can be viewed on modern American paper currency as well as on seashells, and are associated with sea scroll depictions. All of these ocean symbols are prolific on modern American paper currency. The green color shades denote the oceans, fish, sea's and the water environment of the earth. There are reasons for this and later I will address this mater in the context of the Nostradamus prophesy.

When we look for evidence of secret hidden instruments buried in the earth, we must return to the Rosetta stone and the physical and geological origin of this incredible artifact. Just as this ancient stone provided the key to understanding the mysterious Egyptian hieroglyphics, that had been untranslated for thousands of years we begin to realize that because of the stone tablets physical geological origin which is of volcanic origin, the Rosetta stone is trying to tell mankind an incredible story about himself and the earth. Because we now know that the physical nature of this stone tablet has come from a volcano, we must now look into the ancient sacred texts of the Egyptians for further clues, as to the existence of the mysterious hidden instruments of the earth. When we investigate The Egyptian Book Of The Dead, which is the gospel or bible of their religious worship of Sirius, we are shocked to find that throughout this astonishing book, there are references every where to volcanic mountains, lakes of fire, sacred mountains and the cleansing effects that fire and the holy mountains have on the earth. Every where through out this book there are references to Ammen, as well as Amenta which are holy places in the stars to the

Egyptians. There is absolutely no doubt that the volcanoes of the earth, are extremely important to understanding the ancient Egyptian worship of Sirius as well as locating the secret hidden instruments of the earth. When we look at the earth, we are all quite well aware that there are volcanic mountains, lakes of fire and their accompanying earthquake fault zones that run in tandem all over the earth. In any event we now begin to realize that when we look at the entire earth, and we begin to connect man's ancient temple sites, active and inactive volcanic sites all over the earth an astonishing pattern begins to emerge. Because volcanic lava and magma flows have magnetic properties, as verified by scientists and geologists.

The Piri Reis Map

At this juncture in our great journey, we can now make some observations about the Earth and look at it's surface features in a manner that no one has ever done. We are now going to closely peer at the Earth, from the vantage point that the dawn of the space age is now affording us. In the book The Chariots Of The Gods by Erich Von Daniken, there are two photographs on page 78; that were never available to mankind until the advent of the space age. One photograph shows a U.S. Air Force azimuthal equidistant map of the world, with the center of the map being Cairo Egypt. The Air Force azimuthal maps were made possible from the photographs taken by the lunar probes of the 1960's, as they travelled to the moon to do survey work for future manned landings. What occurs in this photograph is that the spacecraft camera lens photographs the Earths surface with Cairo Egypt being the focus of the picture. Everything that is in a radius of 5,000 miles of Cairo becomes increasingly distorted because of the spherical shape of the Earth. The continents away from the center of Cairo sink downward, and turn into the shapes that people have never seen before. This is not a trick or optical illusion, but results from the round shape of the Earth. The point is that the Air Force azimuthal equidistant projection, as well as the American lunar probes and the Apollo 8 lunar flight photographs all show the exact same features of the Earth.

Now we come to the Turkish Admiral Piri Reis. At the beginning of the 18th century, a number of maps belonging to Admiral Piri Reis were discovered at the Topkapi Palace. In addition, two atlas maps which were in safe keeping at the Berlin State library, also came from Piri Reis. These atlas maps depict the exact reproductions of the Mediterranean and the area around the Dead Sea. American cartographer Arlington H. Mallerey was given use of the maps, and together with cartographer Mr. Walters of the U.S. Navy hydrographic bureau they constructed a grid and transfered the maps to a modern globe. Mallerey made a fantistic discovery, when they found them extremely accurate. The Piri Reis maps, matched not only the Mediterranean and the Dead Sea but also accurately portrayed the coasts of North and South America. The maps also showed the exact locations and the outlines of the continents and the interior topography. In addition islands, plateaus, rivers, mountain ranges, and even mountain peaks were all drawn in with great accurancy.

In 1957 the maps were given to cartographer Jesuit Father Lineham of the U.S. Navy and the director of the West on Observatory. After many exhaustive tests Father Lineham also confirmed that the Piri Reis maps were completely accurate. Even more incredible discoveries were to be made when mathematician Richard W. Strachan and professor Charles H. Hapgood researched the maps. The work and research that they did on the Piri Reis maps could never have occurred, until the dawn of the modern space age and until man was able to send space probes to photograph the Earth. They compared the lunar probe photographs of the Earth, and other space photographs taken of the Earth by Apollo 8 and other spacecraft. They made an astonishing discovery when all the photographs matched in exact details, the Piri Reis maps. Their conclusion was that the ancient Piri Reis maps were drawn from aerial photographs of the Earth, which were taken in ancient times! The question is, how could ancient cartographers have flown into space and photographed the Earth; and then flown safely back to the surface. All of these events occurred with spacecraft and photographic equipment, that did not exist at that time! There are answers to that question, but first we must look at what the Piri Reis maps show, because they are identical to the U.S. Air Force azimuthal equidistant projections as well as the lunar and Apollo space photographs of the Earth.

Few people know what an azimuthal equidistant map is outside of cartographers, and most people could probably care less, however the implications of what this kind of a map shows are quite stunning and bizarre. We must now remember what mankind's ancient artifacts tell us about the Gods from Sirius and we must recall what the three orders of the ancient Greek and Roman stone building columns stood for. We must remember that we live on a planet that is drenched in water, and we must remember that scrolls, volutes, rosettes, oceans, salt and other natural Earth materials such as limestone, marble, granite, alabaster, basalt and other geological characteristics of the Earth are all related to. We must remember that all of these elements are related to the oceans, seas, fish, sea shells, ancient Egyptian pyramids, cattle, bulls and a time in man's ancient history typically known as B.C. or before Christ. A time that extends into the future 2,000 years to the year 1999, when our ancient ancestors say that the two great leaders will be friends. President Bill Clinton or (B.C.) and the Russian leader are indeed friends, and the west is helping out this beleaguered leader; thus confirming the Nostradamus prophesy.

We must remember that our human bodies are made of the same calcium as that of sea shells and that our bodies are 70% water based and 30% bone calcium which is the same as the Earths water and land mass composition percentages. We must remember that our ancient ancestors worshipped the star system Sirius as well as the ancient Babylonians, Mespotaminas as well as others all said that Oannes and the superior beings from Sirius are related to the oceans and waters of the Earth. We must remember that our ancient ancestors tell us through their written artifacts that these superior beings are the ones who created our Earth, and built it to their will. We must remember that these beings are The Masters Of The Waters, The Monitors and The Instructors, and there is no getting around the fact that mankind lives on a planet that is drenched in water. We must remember that our blood, perspiration, urine and our very flesh is salty. We must remember that if man does not drink water in 3 days he will die, and we must remember that our ancient ancestors all over the Earth spent thousands of years building incredible pyramids, temples, edifices and all manner of great monuments to the Gods from Sirius. We as humans cannot deny these facts, but must come to terms with them.

The incredible Piri Reis map as well as the U.S. Air Force azimuthal map and the lunar and Apollo space maps all, show the Earth as they have never been viewed and "understood" before. All of these maps show the North American continent including Canada, Alaska and all the surrounding lands. The North American continent is the exact shape of a giant fish, with it's gaping open mouth. The South American continent is the shape of a step pyramid and a fish, bringing to mind the strange "mixed" animals of the Zodiac. The African continent is still the shape of a bulls head, thus the reference to the "horn" of Africa. Antarctica is the remarkable shape of the mysterious volcanic black basalt stone heads, of Easter Island. The Easter Island volcanic statues look remarkabley like the heretic Pharaoh Akhenanton, of the Egyptian Amarna Revolution. As I mentioned earlier, this Pharaoh has been credited with setting the foundation for the emergence of the Christian religion: because he instituted the original idea of worshipping only one God. It is believed by some historians and scholars that the one God worship by Akhenanton, served as the original example and blueprint for the Judeo-Christian religion. History has described him as having a very long and almost distended and unnatural head and body characteristics. He is depicted with his uraei hat or bowling pin shaped Pharaoh's hat on his head. The ancient Piri Reis maps as well as the modern space age photographs all show Antarctica, as the remarkable shape of the black volcanic heads of Easter Island with the Australian landmass appearing as the Egyptian Pharaohs hat. The Australian land mass "hat" appears to be coming down onto the head of the Easter Island head statue, of Antarctica. The truth of the matter is that the stone volcanic heads of Easter Island do in fact have "hats." for their he ads.

The ancient Piri Reis map as well as the U.S. Air Force azimuthal maps, and the American spacecraft photographs all show the astounding physical representations of the superior beings of the star system Sirius. These representations are all the various monuments that have been built or worshipped on the Earth by our ancient ancestors, and they depict all the ancient symbols that are associated with The Masters Of The Waters from Sirius. Further confirming the modern discoveries that the space age is affording mankind, ancient sacred Egyptian religious texts foretold that mankind would reach for the stars and would one day discover the secrets of Sirius.

The Dark Rite

Robert K. Temple the author of The Sirius Mystery, talks about the Trismegistic Literature and The Virgin Of The World.

Quote:

Men will seek out... the inner nature of the holy spaces which no foot may tread, and will chase after them into the height, desiring to observe the nature of the motion of the heaven... These are as yet moderate things. For nothing more remains than Earth's remotest realms; nay, in their daring they will track out Night, the farthest Night of all ...

From The Trismegistic Literature whose author is unknown and is indeterminately old.

Thus we have one small and extremely ancient part of the prophesy, that has correctly foretold that mankind will travel into space and send space probes into our solar system, and beyond to observe the nature of the heavens. The literature is absolutely correct because mankind has already landed men on another world, and he is currently exploring the solar system.

The ancient Trismegistic Literature "Hermetic," as I have already mentioned is hated by modern scholars who say that it is neoplatonist and as such is a mystical work. The fact remains however that this ancient treastie continues to be correct in it's prophesies about mankind and his ventures in the universe. When we examine The Dark Rite and the term Night, as mentioned by The Trismegistic Literature, we find that these terms have their origins rooted deep in human history. These terms refer to the fact than when a person died, his soul or spirit physically travelled to heaven or Sirius through the "Night" or the ink blackness of space. Our ancient ancestors inform us that this was the Dark Rite of that journey through the cold blackness of space. The human custom of wearing black clothes at the funeral of a deceased person, is one that has it's origins in the Middle Eastern countries, and this tradition pre-dates Christianity. Long before Christianity adopted the black clothes custom at the funeral ceremony, mankind had a long history of practicing the Dark Rite clothes ritual. The custom that was adopted by the Christians, is a left over vestige of the religious practises of the ancient peoples of Egypt, Mespotamia, Babylon and other ancient peoples of the Middle East. The Dark Rite and Night are all connected to Isis and the star system Sirius.

Just as our ancient ancestors knew of the Dark Rite that involved the blackness of space and Sirius, modern man is about to find out about his history through the Dark Rite or Night as well by way of the modern space age that we are in now. All of this returns us to the shocking Piri Reis maps and the modern space photographs of the Earth, that show the accurate representations of the ancient sacred symbols of mankind as well as The Masters Of The Waters from Sirius. As I spoke of earlier the entire past history of mankind is bathed in a rich history that was greatly influenced by the Earth's oceans, sea's cows, bulls, volcanoes, as well as sacred pyramids and temples that were built out of limestone and marble which have their origins in the sea.

The Modern Signs Of Sirius On Earth

Because of the great influence that the Egyptian religion exerted on the Greeks and the Roman pantheon of Gods, today Western civilization is awash and is based on many ideals and principles of the ancient Greeks and Romans. For instance, copies of the famous fluted columns of the ancient Greek civilization, can be found on important buildings all over Western civilization. The traditional classic style of man's past, has been duplicated all over the world and in particular our own civilization. Most people except historians, scholars, and architects as well as others do not realize the true meaning, origin, or history; and the implications of the ancient classic style. This very beautiful style has been copied in architecture, sculpture, design, furniture, paper currency as well as other mediums. For instance, in the United States there are numerous government buildings all over the country that were built in the classic style. All one has to do is go down to City Hall and one can view the scrolled capitals on top of the fluted columns, and one can also see the other ocean and fish symbols on these classic buildings. The reason for this, is that it has been a tradition in this country to build important buildings in this style, because our very democracy came from the Greeks. It was therefore natural to copy their beautiful style of architecture. It has also been a tradition because of the close ties between Western civilization, and the civilization of the ancient Greeks and Romans.

In America there are American cities that are named for cities of the ancient world, and yet others named after the famous scrolled land of the Greeks. There is Athens Georgia, Memphis Tennessee, Philadelphia Pennsylvania, Alexandria Virginia as well as other American cities. When we look at the modern American currency, we find that the paper bills are full of the typical classic style of scrolls, volutes, rosettes and the ocean and sea symbols of the Earth.

The world wide ancient symbols of the Egyptians and their worship of Sirius, is depicted on the American one dollar bill. The pyramid has a single eye at the top edifice, and on the left side of the pyramid it is covered with ships rigging. To the ancient Egyptians, the Pharaoh was transported in a large boat to heaven from his pyramid tomb. Why are all these ancient symbols of Egypt on the American one dollar bill? There are reasons for this and it has to do with the Nostradamus prophesy of 1999. However the official explanation from government documents is that, the ancient Egyptian pyramid was placed on the American one dollar bill because the pyramids represent permanency for the American government. The single 'eye was placed at the top of the pyramid in recognition of the fact that, the country is under the watchful eye of an all seeing God. Government documents admit that the green color of the American dollar bill, cannot be explained and no one has any idea why the bills have always been printed in various shades of light green. The truth however lies with the ancient history of mankind as well as the classic style of Greece and Rome. Earlier I said that the Pharaoh traveled to Sirius in a boat, which was intended to take him to heaven or Sirius. The star Sirius, as well as Jupiter are known as the "Eye" stars.

Sirius as we all know now, is known as the Dog star as well as the eye star. Sirius is also a southern constellation magnitude 1 star, and is associated with the Southern Cross constellation. The Southern Cross constellation is the symbol of water and this cross constellation, is the source of where the cross symbol came from. The cross symbol predates Christianity by thousands of years, and is yet another symbol that was adopted by early Christians for their new religion. Water as we all know is a very common element in the universe, therefore a number of additional observations can be made when we look again at the Piri Reis map as well as the modern spacecraft photographs of the Earth. Earlier I discussed the ancient Dogon people of Mali Africa and their worship of the Nommos from Sirius. Their ethnic name means to make one drink. Originally these people lived in the land of Kush, near the fifth cataract of the Nile river. They were ruled by the ancient Egyptians, and their ancestors ironically also ruled the later Egyptians when they became part of the final ruling dynasties of the Egyptian civilization. The ancient Kushite Pharaohs also worshipped the superior beings from Sirius, and built small pyramids and rock tombs in honor and worship of the Sirius Gods. They also cut tombs from the solid granite mountains, in order to expedite the travels of their dead Pharaohs, to the star Sirius. Some of these ancient temple and religious sites can still be seen today in Nubia. For reasons unknown however, a splinter group of the Dogon people broke off from the main group and began to migrate into western Africa, where they eventually settled in Mali Africa. Despite their great travels over thousands of miles and over 2,000 years, (these mysterious people never forgot or abandoned their religious worship of the star Sirius.) Mali is near the western region of African coast and is known to geographers and climateologists as the Sahel region of Africa. Climatelogists now believ that the Sahel region is the birthing source of the gigantic hurricanes that are born in the Atlantic ocean. These huge storms are the ones that are responsible for the devastating hurricanes that pound the Caribbean, and the gulf coast of the United Statss. When we look back at the electromagnetic pyramid images around the Earth, we discover that the Sahel region of Africa lies in side the apex of the giant pyramid image. This is much like the tornados that are formed in Colorado, and then continue to form in the mid west thereby causing other tornadoes elsewhere.

More and more scientific evidence continues to accumulate that confirms that the Sahel region is indeed the source of these hurricanes. It is not known why hurricanes form in the mid Atlantic ocean, but when we look at this location we see that it is right in the middle of the gigantic electromagnetic pyramid image. The ancient Egyptians buried their

Pharaoh right in the middle of the great pyramid at Giza. As these huge hurricanes approach the United States, they produce highly destructive winds and they dump vast quantities of rainwater on their victims. This fact does give great meaning to the Dogon name which means, To Make One Drink! The huge amounts of rain are dumped into the gaping mouth, of the North American fish continent.

When we examine the position of the North American fish continent, and the South American fish step pyramid continent as well as the African bull continent-we then realize that every one of these ancient sacred symbols of our ancestors and Sirius, are facing towards the bottom of the Earth, in the direction of Antarctica. Antarctica is at the bottom of the world, and it's surface features face downwards. Earlier I briefly mentioned the Egyptian heaven or the sacred location called Amenta or the Underworld. This is the final destination of the dead, who are rewarded with eternal life as long as they have lived a just life on Earth. Most people who have ever heard of the Egyptian Underworld have stereotyped it as physically under the Earth, or deep in the bowels of our planet where there are evil demons who perform unspeakable deeds on their victims. Such stereotyping by humans is one of the reasons that different peoples around the world, cannot live in peace. For instance, look at what is happening in Bosnia Hercegovina and the former Yugoslavia. All over the world different people have stereotyped their fellow human heings, and this is immoral; and could well be fatal to people who do this. This kind of thinking often leads to a life of bigotry.

In the case of the Egyptian Underworld, the stereotyping of this location as being an evil place full of dread, is completely wrong. According to the ancient Bible of the Sirius religion, it is heaven and is the location of the star Sirius. Sirius is a binary star system that consists of Sirius A and Sirius B. Sirius A is a magnitude 1 star, while Sirius B is a dense electron degenerate star, and according to the ancient texts there are a number of giant Jupiter sized planets that are orbiting these stars. These planets are the locations where the superior beings live, and is called Amen or Amenta. Sirius and it's southern location, is however visible from July thru September in the eastern skies of the northern hemisphere. Sep is one of the ancient Egyptian Gods of the Sirius religion, and the month of July is the 7th month; that is mentioned in the Nostradamus prophesy. The southern location of Sirius which is physically under the Earth's orbital plane, is the reference to the name the Underworld. Therefore when the ancient Egyptians referred to the sacred Amen, heaven or the Underworld, they were referring to the star system Sirius which is below the Earth.

When we examine the three months, which is a sacred number; from July to September, which is the number 3 and stands for the Earth we discover that an other set of numbers was also sacred. The ancient Egyptians took seventy days to embalm their dead Pharaoh, and this number had great significance in the Sirius religion. The proper procedures had to be strictly followed, in order for the Pharaoh to safely reach his journey through The Black Rite of space to Sirius. In the final analysis we learn that in the ancient Egyptian religious worship of Sirius, the Underworld or Amen was the actual place that the dead journeyed to, when they died. When they travelled through the ink blackness of space or The Black Rite, they were judged according to how they behaved on the Earth. Consequently our ancient ancestors are informing us that the religious worship of Sirius and it's superior beings, are The Masters Of The Waters and the Earth,

What has taken place on the Earth and involves all life on the planet, is that the superior beings or Gods from Sirius, literally terra formed the entire Earth; into land forms and shapes that are symbols of the religious worship of Sirius. Those shapes and symbols that I spoke briefly about earlier, are the representations of the oceans, seas fish, sea shells, scrolls, dog, bull, pyramid, rosettes as well as other symbols of the waters. All of these symbols are easily visible on the ancient Piri Reis maps, as well as the modern spacecraft photographs from Apollo 8 and other space probes; all of which was foretold in the prophesies of the ancient Trismegistic Literature.

The question that we must ask ourselves, is there any hard physical proof that can be presented; that proves that the Earth was indeed terra formed into the ancient sacred symbols of the Sirius religion? How could the Earth have been terra formed by mysterious beings from another star system? How could superior beings from Sirius, who are associated with the waters and are related to amphibians possibly be the fathers of mankind? How could they have left be hind the Monitors known as Bes to watch over mankind?

Up to this juncture of this book, any reasonable person has to agree that the Earth is indeed the water planet, and is literally drenched in water vapor and water. There is no getting around this fact no matter how hard we try, and there is no getting around the fact that our own human bodies are in fact mostly water and calcium based. It is a well known scientific fact that our own Earth as well as Jupiter, are surrounded by powerful magnetic fields. There is no question that when one connects mankind's ancient temple sites, with the volcanic regions of the Earth there is in fact pyramid images all over the Earth, and indeed an amazing amount of tragic events have taken place on these pyramid image boundaries. So the question remains, what hard physical proof exists, that proves that the Earth was terra formed into the Sirius religion symbols? For an answer to that question we must turn to the geology of the Earth.

Daniel Masias (Muh Sigh Uh)

The Terra Formed Earth

In order to present the world wide geological proof, that superior beings from Sirius did in fact terra form the Earth, we must now turn our attention to the geology of our planet. We must go back to the 19th century and investigate the Austrian geologist Eduard Suess. Seuss promoted the radical and absolutely shocking idea to his contemptuous colleagues, that some of the present continents had at one time formed a large southern continent which existed millions of years ago and that at some point in the great distant past for reasons unknown the giant southern continent broke apart. At that time in the 19th century, some of the radical ideas of Seuss had merit because geologists recognized that there existed definite floral and faunal connections between the continents. During his day more and more scientific proof for his new theory came to light, that verified this radical new geologic history of the Earth. Seuss called this large southern super continent, that was supposed to be located some where near present day Antarctica by the name of Gondwanaland. It was not until 1910 and the entrance of Alfred Wegener that a similar but all together different concept of a large southern continent emerged. Wegener who was a meteorologist was highly respected for his work in climatology and paleoclimatology. He developed a new geologic theory called continental drift, which enjoyed much supporting evidence. In the 20th century there was a lot of opposition to Wegener's radical new theory about the existence of an ancient super continent, that encompassed all of the Earths continents into one gigantic land mass called Pangea. For years the debate over the existence of the super continent raged, and it was not until the 1950's and the 1960's that the radical new theory of Wegener began to prevail over his ardent critics. Because of the advent of new sophisticated scientific instruments, which was related to the dawn of the space age; that scientist began the slow patient work of isotopic dating which eventually confirmed the radical new theory of Alfred Wegener. The dawn of the space age further confirmed that all of the continents were joined into one super continent, when space probes were sent into space to photograph the Earth these modern discoveries here spoken of in The Trismegistic Literature, when it says that man will learn the secrets of Sirius.

This shocking new concept was called continental drift and it proved once and for all that at one time millions of years ago, all of the Earths land masses were joined together in a southern location; and that for reasons unknown the Pangea continent began to break up therefore continental drift or the movement of the continents over the hot mantle of the Earth, is now an accepted as well as proven scientific fact. When we look at a globe of the Earth today, if we examine the north and south American continents; and turn the globe at a certain angle we can draw a gigantic world wide S that starts from Gibraltar and then extends through the Americas until it's lower S points right at Easter Island. This is just another of the many world wide physical signs of the superior beings from Sirius.

We now turn our attention to the National Geographic magazine that is dated September 1986 page 390. Here again we have the month of Sep tember, which as we now know there is an Egyptian God named Sep. The magazine researched a 28 page study on meteorites and the Earth. On page 400-1 there is a full page map of the Earth that shows all of the worlds continents and land masses. The purpose of the map is to show the Earth's pockmarked face, and the ancient impact sites all over the world. The map explains that there are 116 meteor impact craters around the Earth, and that only 13 of these sites have definately been attributed to meteors. That leaves a total of 103 other impact craters around the world that through chemical and mineralogical tests identify these sites as probable meteor impact sites. The known 13 meteor sites around the Earth have been confirmed by chemical and mineralogical fingerprints, as well as the recovery of meteor material at these impact sites. Of the 103 impact sites around the Earth, no meteor material can be found but the gigantic sites are still being counted as meteor craters anyway! The reason for this is that these huge impact sites are exhibiting certain mineralogical and chemical finger prints, that the 13 known meteor sites are showing. Geologically chemical and mineralogical finger prints leave certain trace chemicals; and the surrounding rock strata leave certain shock patterns. However the most interesting fact concerning the other 103 world wide craters, is that there is absolutely no meteor ejecta to be found anywhere at these world wide crater sites Because of the massive size of some of these so called impact meteor sites, which in some cases are 100 miles in diameter geologists should be able to easily recover the meteor ejecta. For instance if one travels to Meteor Crater in Arizona, which is just under one mile in diameter one can recover meteor ejecta from this relatively small impact crater site. This site is one of the confirmed impact crater locations of this study. Any reasonable person who looks at the data and sees that 103 gigantic meteor impact sites did not leave any meteor ejecta material, and yet a tiny meteor site like Meteor Crater in Arizona did has too agree that maybe the 103 world wide crater sites are not meteor impact sites but something else!

This is where our ancient ancestors and the superior beings from Sirius come into the picture. Before we can thoroughly examine their involvement in the terra forming of the Earth, we must first examine another geological mystery that scientists are at a loss to explain.

The meteor crater map in the National Geographic magazine also, reveals to us another enigma, that has caused a great debate among geologists around the world. Scattered over the Earth, there exists three enormous fields of what scientists call glass tektites. Geologists and scientists have not been able to reconcile how these billions of tons of glass objects, fit into the Earths geologic history. Two of the world wide tektite fields are shaped like rounded pyramid shapes and the third one is the shape of an oval. The great mystery and the problem that is associated with these enormous fields of glass tektites, is that they cannot be explained by any of the Earth's known geologic processes Scientist do not know where these enigmatic glass objects came from, or precisely how they fit into the geologic process of cause and effect. The best explanation that the geologists and the scientific community can come up with, to explain the billions of tons of tektites is that they were splashed from giant meteorite impacts with the Earth. This is a nice neat explanation for the glassy objects, but when we look at the world map; the tektite fields are located in only 3 confirmed and 20 probable meteor impact sites around the world. That leaves a total of 93 other crater sites that are either hundreds or thousands of miles from the three tektite fields. The fact that the tektites are not near the 93 impact crater sites, also tells us what scientists have known all along There is a very big mystery concerning these glassy objects, and nobody has been able to explain where they came from. Returning to the 116 impact sites all over the Earth, we naturally have to include Alfred Wegener and his accepted discovery of the Pangea super continent and continental drift. As I discussed earlier, Wegener postulated that the Earth's entire continents at one time were all joined into one enormous super continent; millions of years ago. When we closely examine the National Geographic map, a very interesting pattern begins to emerge. Upon further examinations of the different phases of the break up of the Pangea continent, we begin to see a pattern emerge. When we look closely at the meteor impact crater sites on the National Geographic map, we can then see that the crater sites are located in or near the continental coastal areas where the continent of Pangea broke and then separated into it's respective continents! It does not take a genius to see that a great deal of the meteor impact sites are located at the "breaking points," where the continents broke apart or were scored by some unknown process! What do I mean when I say scored? Anyone who has ever cut glass or some type of tile will know right away what scored means. When a person cuts a piece of glass, we must first draw a straight magic marker line, on the glass that we want to cut. With the glass cutter we then carefully score the surface of the glass along the marked line, that is we gently but firmly use the tiny round blade of the glass cutter to break the surface of the glass; along the marked line. The glass cutter will just break or score the surface of the glass, so that all you have to do is turn the glass over and elevate one end about 1 inch. We then give the glass a quick karate chop with an open hand, and if we scored the glass properly the glass will break perfectly along the marked line.

Ancient Egyptian sacred texts and mankind's ancestors all over the Earth spoke of the Gods from the skies, who created mankind and the Earth. Now that evidence is revealing itself through the space age. It is through the use of space probes and satellites that were equipped with cameras and other devices, that the accuracy of the National Geographic map is able to show the various crater sites all around the world. When we look at the enigmatic glass tektites and the enormous fields that they comprise, we are able to determine that the tektite fields are located in the softer geologic strata of the Earth. It is in these same areas that the Pangea continent broke apart. By the same token, when we look at the location of the world wide gigantic crater sites we learn that the geologic strata of those areas are generally a harder material. There is no doubt that the superior beings from Sirius literally created and terra formed the Earth's continents with some sort of enormous crater forming devices The gigantic explosions were needed for the harder Earth strata, in order to score the Earth's surface and then allow the continental drift forces to break and then separate the Pangea continent into the shapes that we see today. When we examine the glassy tektites and the world wide fields that they are located in, we can see that they are located in the softer Earth strata. With some sort of unknown devices, the beings from Sirius used incredible heat vitrification, to literally burn and cut the softer Earth strata to achieve the finer shapes of the continents.

The Enigma Of Meteor Craters

Now we can return to a brief statement that I made earlier in the book, and which I promised to address. Scientists have determined that the Earth and Jupiter are literally magnets, because they have very powerful magnetic fields around these planets. As kids we all played with magnets and we remember that as we ran the magnets through the dirt, we could always attract metal shavings from the dirt. If we compare the Earth, to the magnets that we ran through the dirt as kids then we would expect the Earth to be bombarded by iron meteors in far greater numbers than it is being struck by these space travelers. Since the Earth is such a powerful magnetic force in the solar system, it should be attracting large iron and nickel meteors; on the order and size that struck in Arizona. Scientists acknowledge that they can only find 13 conf firmed meteor crater sites around the world, and when we compare this to the vast numbers of craters on the Moon we would expect the Earth to have many more meteor craters on it's surface. We would expect this, despite the fact that the Earth's weather does erode geological features on the surface. However the eroding effects of the Earth's weather does not go far enough in explaining the lack of meteor craters around the world. The reason that I say this is that mankind has not been able to actually view the process, from the time that a meteor impacts the Earth's surface to the complete erasure of that crater. When we examine the craters of the world, Moon, and the other planets of, our solar system, we notice one fact that we have always taken for granted. When we investigate the meteor craters on planets and their satellites all over our solar system, we find that the vast majority of these meteor crater impact sites were made or appear to have been neatly created. The craters appear to have been made when a meteor struck the surface of the planets in a near perfect frontal impact, in relation to the surface; thereby creating a perfectly round and symmetrical shape. When we investigate the planets and the moon's of our solar system, there does not appear to be a single solar body that has been struck by a meteor that crashed into the surfaces at an oblique angle. A meteor impacting a planets surface at an oblique angle would not create a nice round crater, but would more likely create a long gouging trench like impact site. In almost no instances, there are no goughing like impact sites on the planets and the moons; and my question is why? This is a very odd curiosity because most of the craters in our solar system are all nice neat round impact sites, that have well defined edges. The craters almost appear to have the clean neat symmetry and mystery, as the crop circles of the Earth.

In order for craters to be formed with such neat and orderly defined boun daries, the meteor would have had to have struck the surface of the planet at a near perfect angle to it's surface. This means that the vast majority of round symmetrical craters in the solar system were, struck at near perfect angles to their surfaces. This seems highly unusual, in light of the fact that there is photographic evidence that shows that meteors do in fact approach planets at oblique angles. There are elements at work here that suggests that possibly our ancient ancestors knew more about Earth than we suspect.

On June 30, 1908 the time frame that Sirius begins the 90 days of it's helical rising in our eastern sky the skies over Tunguska Siberia became an enormous caldron of exploding fire, heat and energy that has never occurred before in recorded time. It is generally believed that a large meteor impacted the Earth's atmosphere, and was somehow destroyed as it penetrated the upper regions of the Earth's skies. The terrible results were that virtually every tree within 20 miles of the Tunguska site were destroyed, and people 40 miles away from the center of the explosion were badly burned by the searing heat of the blast. Horses and incredible 400 hundred miles away, were blown off their feet. The Tunguska explosion has also been attributed to the mid air explosion of a comet, because of the enormous destruction that it caused. On succeeding nights after the blast newspapers could be read at midnight all over England and parts of Europe. This was attributed to scattered meteor dust in the Earth's atmosphere, due to the explosion.

Now we can turn our attention to the location of the Tunguska explosion site. You have probably already guessed where it occurred, and you are right. We will recall that the third great invisible electromagnetic pyramid image, begins at the ancient temple and volcanic sites of Italy, and then travels through the southern Mediterranean; with it's Greek temples and then continues onto Turkey. Iran, and then India. From here the giant pyramid image then travels up into Russia and right through the site of the gigantic Tunguska explosion site! Now we can turn our attention to another astonishing meteor event of the 20th century. On August 10,1972, again just as during the Tunguska explosion which occurred during the time frame of the appearance of Sirius in our eastern skies this meteor incident also occurred during the 90 days that Sirius makes it's helical rising in our eastern sky. This event was documented in the September 1986 issue of National Geographic magazine. By a good stroke of luck, an amateur photographer at Jackson Lake Wyoming was able to take a picture of a 1,000 ton meteor as it streaked across the Wyoming skies, which is the exact location of the fifth giant electromagnetic pyramid image in the United States. The white hot meteor could be seen from Utah to Alberta Canada. According to calculations, the meteor streaked in from the southern hemisphere below Mexico. This location of the world, is where the enormous base of the number one electromagnetic pyramid image is located. From this site the huge meteor flew through northern Mexico, and then entered the air space of New Mexico. At this point the streaking meteor then entered Colorado, which is the apex of the large fifth electromagnetic pyramid image in this part of the world. It then continued up towards Wyoming and flew past the final leg of the fifth pyramid image, and continued up the Earth until it streaked back into space above Alberta Canada. This photograph of the 1,000 ton meteor is proof that meteors do approach planets at oblique angles. If any of them were to impact the surface, they would cause enormous damage on the Earth and they would create a gouging trench like effect. This would be in contrast to the nice neat craters, that we see all over the solar system. Because of these two incredible events in Tunguska Siberia and in Wyoming, there is strong evidence that the electromagnetic pyramid images of the Earth, have the ability to repel errant meteors. In addition both of these startling events took place during the 90 days that are known as The Dog Days of summer, which is in reference to The Dog Star Sirius. The 90 day time frame of the hottest and muggiest days of the summer, that is referred to here is from July to September. In The Sirius Mystery by Robert Temple, he postulates that these superior amphibian beings who live on a gigantic Jupiter sized water planet would almost certainly live in or near the water. This is because Sirius A is so big and hot, therefore their watery planet is probably very hot and muggy, much like this planet was when the amphibians and dinosaurs were inhabiting the Earth. Therefore these beings would most likely live in cities that were in reed like environments that were very warm and humid, with lots of water for protection against the powerful solar and ultraviolet rays of Sirius A. They would need protection from the damaging rays of the sun, therefore they would seek that refuge in underwater cities which makes complete sense when we realize what man does here on the Earth. When we look at the nuclear energy industries around the world, these nuclear power plants are built with water pools in them, to place the nuclear fuel rods into. This is done to keep the rods cool and in fact the water pools also protect the plant workers from the harmful effects of the nuclear fuel rods. Water is also used as a cooling agent in these plants as well. The current thinking now in the space community is that when astronauts are sent to Mars, the spacecraft will have to be encased in some sort of water environment that would be contained in an outer hull or shell. The purpose of this is that the astronauts will encounter deadly ultraviolet rays and solar bursts from the sun, that could kill them. It would be impractical to shield the spacecraft with lead, so the best and most economical way to protect the astronauts is to shield them with an outer hull that contains water. It is a well known fact that water affords the best protection against deadly radiation.

When we go back to the ancient Dogon people, and we re-examine their words; about the day when the Nommos arrived on Earth we remember that they said that the Nommos or Oannes, had to be in the waters. When their spacecraft landed it was placed in a dry hollow in the ground. The hollow then filled up with water from the spacecraft! Thus these ancient Dogon people are actually describing a process that is used by the nuclear energy industry all over the world, of using water as a shield from harmful solar rays. Other ancient peoples who came in contact with Oannes all describe these beings as part amphibian and they resided in the Red Sea at night, this was their abode. When we look at the geological history of the Earth we learn that at one time the planet was inhabited by amphibians and dinosaurs, in a world that was hot, muggy and humid which is a good description of the planets of the Sirius system. Our ancient ancestors have left written artifacts that inform us that at one time here on Earth, there was a watery abyss where there resided hideous beasts. Can these hideous beasts be the dinosaurs of millions of years ago? We are also told that they were destroyed by someone, and for reasons unknown. Geologists now firmly believe that a gigantic meteor impacted the Earth some 70 million years ago and caused the extinction of the dinosaurs as well as most life on this planet. Could the superior beings from Sirius have caused the extinction of the dinosaurs here on the Earth? In any event all of the elements from the use of water to protect life from harmful radiation to the hot muggy days from July to September, known as the Sirius Dog Days of summer; do in fact correspond to the ancient written words of our ancestors that the hideous beasts of the watery abyss were inhabiting the Earth.

In examining the possibility that meteors have caused great destruction here on Earth, we are struck by the fact that from the time of the Tunguska explosion to the 1972 meteor incident in Wyoming a short

64 years had passed The time frame of 64 years is very short, when compared to the geologic record which measures in the millions of years this short time frame of 64 years, brings us back to my original statement that the Earth should be the object of a good deal of bombardment by meteors, because it is described as a literal magnet just like the planet Jupiter. The religious worship of Sirius and the Nommos began during the reign of the ancient Egyptians. They tell us that the Nommos imparted knowledge to them about the giant planet Jupiter, Saturn and their moons. The superior beings from Sirius told the Dogon people, that some time in the future they would return to the Earth to rule mankind and their Earth.

As I spoke earlier, but bears some repeating at this juncture the ancient Egyptian religious worship involved Ra the sun God. Osiris and Ra were always associated with flight and the Divine Eye or Eye Stars, which were Sirius and Jupiter. Throughout every religious depiction that the ancient Egyptians painted or carved, there has always been one recurring theme that never goes away. I am speaking of the single very pronounced Eye that we see on the face of Osiris, Horus, Anubis, Isis, Amen Ra and every other Sirius God. This fact is very important because it involves the giant planet Jupiter, and the astonishing events that are going on there! Jupiter is the fifth planet from the sun and it's number corresponds to the sacred number five, which is also related to the electromagnetic pyramid images of the Earth. The most important moon of Jupiter is Io, with it's entire surface of active volcanoes and lakes of molten volcanic fire. These lakes of fire are spoken of in the ancient sacred texts of the Sirius religion.

The fifth planet corresponds to the fifth civilization of mankind, which our ancestors say we are now in.

Earlier I promised to return to a discussion of Jupiter and it's astonishing moon Io. Besides the Earth, Jupiters moon Io and the Earth are the only known celestial bodies in the solar system, that exhibit active volcanic eruptions. The ancient Egyptians, Dogon, Greeks, Romans, and other peoples all worshipped Jupiter. What is shocking to the scientific community is that scientists have discovered that as Jupiter rotates on it's axis, at an astonishing rate of 9 hours and 50 minutes; after a period of time and when it's volcanic moon Io is in a certain position relative to Jupiter-suddenly the gigantic planet starts to furiously begin transmitting extremely powerful electromagnetic waves at the Earth! The planet does not transmit it's powerful electromagnetic waves at Mercury, Venus Mars or any of her planet in our solar system. It only transmits these enormous waves at the Earth, and to no other planet! This is an astonishing event that was discovered in the early 1950's with the advent of radio telescopes, and as a result the scientific community of the entire world has taken notice of this fact. The discovery that scientists made back then is related to the ancient prophesies, that mankind would learn the secrets of Sirius. Because of the powerul waves of electromagnetic energy that are beamed at the Earth, Jupiter and it's moon Io have gained the undivided attention of scientists around the world. As a result of this scientific fact, a total of 4 American space craft have been urgently dispatched to Jupiter and Io, to investigate the powerful source of the electromagnetic waves on Jupiter. Pioneer 10 was launched on March 3,1972: Pioneer 11 was launched on April 4,1973: Voyager 2 was launched on August 20.1977: Voyager 1 was launched on September 5,1977. Every one of these spacecraft were launched towards Jupiter, in order to investigate the powerful electromagnetic waves that are being sent to Earth! At this very moment there is another American spacecraft that is orbiting Jupiter, to again investigate why this giant planet is transmitting powerful electromagnetic waves at the Earth! The spacecraft that was dispatched to Jupiter is called Galileo, and it has been at Jupiter since December 1995. The Galileo spacecraft is by far the most expensive ever flown and it's total cost is around 1 billion dollars. The price tag of this spacecraft is well worth it's cost, because it is by far the most sophisticated probe ever built by man and it also has the most important mission to perform for mankind. This spacecraft is actually two probes in one, because when the Galileo craft arrived at Jupiter, it was programmed to drop another smaller spacecraft into the clouds of Jupiter; and investigate and identify the source of the powerful electromagnetic waves that are being furiously beamed at the Earth!

Just like the movie "2001-A Space Odyssey." NASA has urgently dispatched a whole series of spacecrafts to this gigantic planet. The current Galileo spacecraft is by far the most sophisticated probe to investigate this planet, and it is also the fifth spacecraft to be sent to Jupiter and it will discover the secrets of Jupiter which is related to Sirius according to the ancient Trismegistic Literature.

In the late 1960's when the classic movie "2001 A Space Odyssey," was filmed, there were people in the motion picture industry who learned of the astonishing events that are taking place between Jupiter and the Earth. It was someone's opinion that Jupiter has some sort of intelligent life on it's surface, who are responsible for the gigantic electromagnetic waves that are being furiously beamed at the Earth! Because of what the scientific community had learned about Jupiter when the radio telescope was invented, the classic movie "2001 A Space Odyssey," was therefore filmed to document this astonishing discovery. It is no accident that the black monoliths in the movie, look like the black volcanic basalt Rosetta Stone. The monoliths are of course much larger, than the real black Rosetta Stone which is the real life symbol of Sirius and

Jupiter. In the real life story of the astonishing Jupiter mystery, the black monolith is located on the Earth and on the moon; and it is a volcanic basalt stone that is 3.75 feet long by 2.5 feet wide the exact sacred numbers of our ancient ancestors. It is with the Rosetta Stone that the true history of mankind and the Earth is revealed, and it identifies the superior beings from the star system Sirius.

Journey To The Secrets Of The Solar System

Like the Galileo spacecraft, there are other highly advanced space probes on their way to investigate other signs of intelligent life in our solar system Least we forget about other astonishing discoveries, made at Saturn by the Voyager spacecraft; we remember some facts that now make sense. During November of 1980 the Voyager 1 space probe reached Saturn and while there, it discovered some very mysterious facts about the rings of this giant planet. For instance there are "shepard moons" that actually defy the known laws of gravity, as Sir Issac Newton understood them. Newton would have been pulling out his hair, in attempting to understand how these "shepard moons" are able to defy his laws of gravitation and planetary motion. These moons actually "herd" back into bounds, rocks and boulders that try to escape the F ring! This is much like a sheep dog herding the stray sheep back into the main flock, except the dog is defying the laws of gravity and Sir Issac Newton. Saturn also has mysterious "spokes" that emerge from the shaded side of the planet, and they burst out into fives" and revolve with the rings. these spokes also defy the laws of gravity and they should behave like a satellite, but they don't! Another astonishing discovery that the scientists made concerned the F ring of Saturn. When they looked at the F ring, the ring was split into three strands; and two of them appeared to be intertwined with each other. The incredible conclusion was that these strands looked like a DNA double helix! This braiding effect goes completely against the laws of orbital mechanics for several reasons, and defies the laws of gravity. Needless to say the scientist who were in charge of the Voyager spacecraft were very frustrated, because the information that they were receiving from Saturn did not conform to the known laws of physics and gravity. Everything that they had learned about gravity and Sir Issac Newton, in their universities and colleges seemed to be failing them when it came to understanding the mysteries of Saturn.

The superior beings from Sirius left clues all over the solar system, for mankind to find when he begins to chase out into the height "space." Those clues are the discoveries that have so baffled and frustrated the scientist at Jet Propulsion Laboratories, who were in charge of the Voyager deep space probe. Consequently we see that the ancient Trismegistic Literature was completely correct, in it's predictions that man would chase out into the height and observe the motions of the heaven. On September 25, 1992 NASA launched the Mars Observer which will arrived at Mars in the summer of 1993. It arrived at this planet at the end of August in 1993. It is no secret that Mars has been studied to death by numerous preceding spacecraft, as well as the Viking spacecraft that landed on the surface, and are now inactive. The Mars Observer is the most expensive and sophisticated spacecraft to visit the red planet. It is equipped with the most powerful cameras and imaging devices, so that it can peer down at the surface and photograph areas of Mars that some scientists believe are the locations of pyramid and building ruins! The Viking space probes of the mid 1970' s photographed what appears to be a large "face" on the surface of Mars, and the sole mission of the Mars Observer is to re-photograph these sites. The space agency however will not admit this publicly. Because Mars has been studied time and again, the mere fact that a very expensive spacecraft that is equipped with the most powerful imaging cameras; tells us that NASA also believes that there is a good chance that the powerful cameras will spot the pyramids and building ruins on Mars. Further enforcing this argument is the fact that the space agency does not have the money to waste, in this era of very tight budgets so the Mars Observer is considered a good risk. Otherwise they would not waste their precious resources on a planet that has been studied time and again!

American spacecraft are not the only space travellers to witness mysterious celestial objects. In 1988 two Russian Phobos spacecraft were sent to investigate Mars and the two moons of this planet. Of particular interest to the Russians was the Martian moon Phobos, which has exhibited the most peculiar behavior. Scientific observations and measurements of this small moon, has indicated that it is hollow and possibly under some sort of control.

When the first Phobos spacecraft entered it's Martian orbit, the craft began to study the Martian moon. As the Russian craft approached the small moon, and prepared to probe it with an onboard laser it's cameras began to take video footage of an approaching cigar shaped UFO. The Russian spacecraft was equipped with instruments that would determine if the Phobos moon was hollow! Shortly after the approaching UFO was detected the Russian spacecraft went dead. Later on "Sightings," an American television program, a Russian scientist in charge of the Phobos space mission said that they believed that the UFO destroyed their spacecrafts.

We have barely begun to scratch the surface of the ancient connection between Sirius and mankind. The efforts to discover "Night" is now beginning to assume a universal effort, that involves humans all over the Earth. Not only have American and Russian space probes been involved in this effort, but now the Europeans have joined the American efforts in a joint venture that is called Ulysses. The cooperation between NASA and the European Space Agency has resulted in the dispatching of a spacecraft to Jupiter. On February 8,1992 the deep space probe flew past Jupiter and began studying the planet. The purpose of this probe was to determine why Jupiter is transmitting electromagnetic signals to another star system The Ulysses spacecraft then headed for our sun, to study it.

There is absolutely no doubt that extremely expensive and sophisticated spacecraft have been launched by a number of countries, in efforts to observe and identify a number of very baffling mysteries on Jupiter and Mars. The fact of the matter is, that all of the deep space probes of the last 25 years have been launched in efforts to locate, identify, and photograph the mysterious sites on other planets; that are exhibiting mysterious characteristics. Scientist and space agencies will not admit it publicly, but they know that in our solar system a number of very strange situations exist and their space probes have been involved in the deciphering of those clues. Because of this situation the entire S.E.T.I. program is now underway, in an attempt to detect the presence of other life in the universe. This program is a 100 million dollar and ten year program, that involves a world wide network of giant radio telescopes. This search program is massive in it's scope and it tells us how serious science is in it's efforts; to cull the secrets of the universe.

Daniel Masias (Muh Sigh Uh)

Mother Earth

All of the dilligent efforts directed at Jupiter remind us of the ancient Dogon people, and their knowledge of Jupiter and Saturn; and the moons of these planets. We recall that the ancient Greeks and Romans had a great deal of worship for this giant planet. The Greeks referred to it as Zeus and the Romans called it Jupiter. This planet represented the King of the Gods, and many beautiful temples were built in honor of this God. Zeus was usually associated with thunderbolts and great power over the other lesser Gods, as well as man and the Earth. This God was feared and yet he was loved and revered by our ancient ancestors, who associated him with thunderbolts and lightning.

How ironic it is that every single American spacecraft that has been sent to investigate Jupiter, has in fact discovered that Jupiter does in fact have enormously powerful lightning and thunderbolts that endlessly explode through it's clouds! This scientific fact confirms the knowledge of our ancient ancestors!

According to the ancient Dogon people, the Nommos told them secret scientific facts about their home at the Sirius star system as well as facts about Jupiter and Saturn. As I briefly mentioned earlier, astronomers were not able to confirm certain scientific facts about Sirius until 1970. However these facts were already known for thousands of years by the neolithic Dogon! Again I don't want to belabor the point but, How did the primitive Dogon people who live in caves and straw huts and without the aid of any technology, science, engineering, telescopes or any kind of modern civilization-gain correct scientific knowledge about Jupiter, Saturn, and the star system Sirius, which is 9 light years from the Earth! There is only one plausible and true answer to this question.

That answer is that the Dogon people as well as our ancient ancestors are all telling us the truth about the Gods from Sirius!

When we look all over the world and at our own human bodies, the signs of The Masters Of The Waters, The Monitors, The Instructors, are every where. The Earth is drenched in water and the atmosphere is full of water vapor, which we see as clouds and other elements. The oceans of the world act as carbon dioxide sponges, and purge the pollutants from our atmosphere so that life can continue to exist. This very process as well as other agents has over millions of years, resulted in the geologic creation of the Earth's limestone, sandstone, and marble. Ironically these very building materials were the ones that our ancient ancestors used to build their great pyramids, temples, edifices and other great structures in worship of the Gods from Sirius The Masters Of The Waters. All over the Earth there exists the volcanic mountains and earth quake zones, that when connected to mankinds world wide temple sites form the electromagnetic pyramid images of the Earth because basalt is magnetic. When we look at the Earth's great mountain ranges, we discover that granite mountains are not the majestic smoking behemoths of the classic true volcano. However these enormous majestic granite mountain ranges all over the Earth, were all born of the same source under the Earth! That birthing mother was magma inside the hot bowels of the Earth. Therefore we can see why our ancient ancestors, worshipped mother Earth, father sky, and sacred mountains. One of the most important ancient symbols of man was the tree. There are, what is called the Tree Codes in the ancient world. The ancient sacred tree was in reference to the tree of life, which travels back in time to the beginning of the great pyramids. When the Bible was being written, the Tree Codes as well as the pyramids were already ancient by comparison.

Daniel Masias (Muh Sigh Uh)

In today's modern world, the great importance of trees to mankind's world, is not lost to our sensibilities. Many people the world over abhor the great loss that we are suffering as well as the Earth, with the loss of the rain forest and other habitats of our planet. Our ancient ancestors referred to the sacred tree of life because their physical shape was in the form of the sacred mountain and pyramid, which are the symbols of the Sirius religion.

Practically every facet of the human condition on this Earth is ruled by the pyramid or triangle shape. For instance, the civic, economic, and military power of nations and their standing among one another is governed by the pyramid image. At the very top of the mountain or pyramid image is the superpower, which as we all know is the United States. Below the United States, there are the less powerful countries in descending order. This very same governing pyramid of human endeavor, is the ruling law that affects everyone. This unspoken law of nature is prevalent in business, as well as in our various personal efforts that cover all the ranges of human efforts. For instance the personal efforts that we put into our jobs, families, relationships, leisure time, bad habits, sex lives and every other activity that we as humans are involved in. Topping the tree pyramid are the people who voluntarily put in extra time without pay at their jobs, and below are the rest of the workers who punch the time card, and just put in their eight hours. We all know who we are talking about. This very same situation is also prevalent in every other human endeavor. We are all familiar with this pyramid law and we know it in several forms. Everyone would like to drive a luxury automobile, but not everyone can afford one so the rest of us drive less and less expensive automobiles. This same example can be used for the clothes we buy to the homes we live in to the restaurants that we dine out in. Likewise, every man does not get married to Miss America, so we therefore realize that the vast majority of our human species is in the descending realms of the pyramid law. The law of the pyramid is in every aspect of human life. It is very simple in it's scope. At the top of the pyramid are those who are the best, richest, most powerful, luckiest, as a well as any other number of attributes that are accorded to human beings and their institutions.

When we examine our own human bodies and the very elements that they are made of, we are not surprised to learn that the majority of it is composed of water. Our very bones that form the interior frame work of our. bodies, are made of calcium; which is quite similar to the sea shells of the Earth's oceans All of the various body fluids of the human being is salty in nature. Our tears, urine, sweat, blood and our very flesh, is all quite salty. We know that if a human being does not drink water in three days, he will most certaintley die of thirst. All of these elements of the human body are related to the oceans and the seas of the Earth. Because of this fact, we are all tied to the waters all around us.

When we look closer at the Earth, we discover that there are five different food chains in the world. All of these food chains are pyramid in shape. The first pyramid food chain is located in the sea. The second pyramid food chain involves the forest and it's creatures. The third pyramid food chain involves the Arctic regions of the world. The fourth pyramid food chain involves the Earth's great carnivorous beasts. The fifth and final pyramid food chain involves mankind and the domesticated animals that he consumes. There is quite possibly a sixth pyramid chain, but it is so upsetting that I am not going to delve into this topic and I will make no further mention of this topic other than to say that it has to do with evil, which the number 6 is associated with.

Everywhere on the surface of our planet, there are the signs of the superior beings from Sirius. Hurricanes and cyclones which are known for their profuse winds and water elements, are also known for their

"Eye." The "Eye" of course is very pivotal in understanding the ancient Sirius religion. When we investigate where these hurricanes and cyclones as well as tornados, occur on the surface of the Earth; we then discover that they are formed inside or near three of the giant electromagnetic pyramid images, around the world!

The ancient Egyptians said in their religious texts that "Light" was sacred, and was associated with The Seven Shinning Ones. In reality the blazing white light from our sun was sacred to them and their worship of Sirius. The next logical question that we must ask is why? Just what is sunlight and how can it be analyzed? When we analyze the sunlight that falls on the Earth with a color spectrum, the various colors that we see are purple, blue, green, yellow, and red. These five colors of the color spectrum, that emerge from the sunlight of our sun are again representative of the ancient sacred number 5. When we think back and jog our memory about the stories that we have heard over the years about flying disks and UFO's, we then remember that these five colors are the exact colors that people have reported seeing on the mysterious flying saucers! these five colors are also quite prolific on the ancient tombs as well as their papyrus scrolls, even more so. Many of the Egyptian sacred texts have these colors on their depictions of their religious ceremonies.

In an article by Carl Sagan, who was an astronomer that is dated March 7, 1993; Sagan makes the argument that alien abductions are the sheer figment of peoples imagination. He states his skeptical beliefs that there is no such a thing as aliens and that they are not performing biological examinations of human beings. Ancient Egyptian texts tell us that Bes is the protector of women in child birth, and that these beings are the peoples protector and God. They are also the Monitors of mankind and the Earth.

When we look at mother Earth and mankind, we know that the gestation period of the human being is 9 months. During that nine months, the child is in a complete water environment, thoroughly confirming our ancient ties to the Gods from Sirius, who are known as The Masters Of The Waters. The nine months that are subtracted from one full year, leaves a remainder of 3 months which is an ancient sacred number-and represents the Earth and the dominant species on this planet which is mankind. As I mentioned earlier but bears repeating, the number 9 is a sacred number and it stands for the nine planets of our solar system. The ninth planet or Pluto is the only planet in our solar system whose orbit is very different from all the other planets. As this planet orbits our sun, it rises above and falls below the plane of all the other planets of our solar system. All of the planets in our system orbit the sun in the same exact plane. Pluto is very different, and it behaves in it's orbit in the same manner that Sirius does when the Dog Star rises for three months in our eastern sky, and then falls back into the "Underworld," during the rest of the year. Since Pluto is the 9th planet of our solar system, the comparison to Sirius as being 9 light years from the Earth is therefore unmistakable.

The number 9 was also the sacred number that identified the nine great Gods of the Egyptian Ennead. We now know that our ancient ancestors considered the numbers 2,3,5,7 and 9 to be very sacred, powerful and lucky numbers in their lives. When we take these 5 numbers and investigate them even further, we are able to discover that they are also the exact five numbers of the temperature of the universe! The universe is 2.735 degrees above absolute zero.

Scientists around the world are now discovering astonishing new facts about the universe that we live in. These shocking new discoveries were never postulated by any new theories, and that is why the old theories about the universe are being tossed out the window. It is coming as an incredible rude awakening, that the universe is nothing like we thought it was. Scientists with the use of new and more powerful instruments have begun to discover that the universe if full of invisible dark matter! They are also beginning to discover that the universe is totally different in matter composition, than previously thought! With the help of new scientific instruments, astronomers have discovered that because of the gravitational behavior of galaxies and other bodies in the universe the vast majority of the matter in the universe is composed of invisible dark matter that is not understood by astronomers! The new discovery of this vast invisible material has come as a complete shock to the scientific community; because it was never predicted by any scientific theory or theorist.

Quantum physics as it is now understood, is now postulating that matter can suddenly appear from nothing; much like the invisible matter that scientists have now discovered. What scientists are now saying is that when we look out at the universe at night, or through a telescope what we are seeing in the white stars is vastly out numbered by the invisible dark matter that is everywhere in the universes because the new science says that matter can suddenly be created or exist from nothing or empty space, scientists are examining the dark matter of the universe; in light of the new theories. The new discoveries about the universe are mind bending, however in the Bible or Gospel of the Sirius religion The Egyptian Book Of The Dead has many references to the fact that the Gods were "born from nothing." In this astonishing book, the ancient Egyptians describe the exact same elements of dark unseen invisible matter that the Gods were born from. They describe how their Gods were suddenly born from nothing but empty space, and how they gave birth to themselves! The type of descriptions that the ancient Egyptians speak about in their Sirius religion, are the same descriptions that modern astronomers are attributing to the newly discovered invisible dark matter, that makes up 95% of the universe.

OANNES: Man The Animal

When we examine the powerful and seemingly unsolvable problems that divide peoples all over the world, we wonder what it is about the human mind and the brain; that so divides so many people? All over the world there are conflicts between men and women in one form or another. The Earth is full of different ethnic groups, who are in conflict with other racial groups of people. We wonder if the human race can ever live in harmony with his fellow human beings Somewhere there has to be an answer to all of the racial and ethnic problems that plague man. There are reasons why mankind acts the way that he does, and there exists ancient written records that were authored by Oannes the superior being from Sirius. The ancient words of Oannes are indeterminately old, and from all accounts his words pre-date the pyramids. His ancient words come down to modern man by way of Berossus, of which I spoke of before.

Berossus was an ancient Babylonian priest who lived at Babylonia which was located between the Tigris and Euphrates river. Berossus as well as Apollodorus wrote an account of an intelligent being called Oannes, who emerged from the Red Sea; and gave civilization to mankind. Oannes who has been deliberately ignored and shunned by past and present traditional scholars, wrote these words about mankind; despite the fact that efforts were made to destroy his writings. Very little of the writings of Oannes survives to this day, the following written description about the "condition" of mankind in extremely ancient times is given by this being from the star system Sirius.

Quote:

    There was a time in which there was nothing but darkness and an abyss of waters, wherein resided most hideous beings, which were produced of a two fold principle. Men appeared with two wings, some with four and with two faces. They had one body but two heads the one of a man, the other of a woman. They were likewise in their several organs both male and female. Other human figures were to be seen with the legs and horns of goats. Some had horses feet others had the limbs of a horse behind, but before were fashioned like men, resembling Hippocentaurs. Bulls likewise bred there with the heads of men and dogs with fourfold bodies, and the tails of fishes. Also horses with the heads of dogs Men too and other animals, with the heads and the bodies of horses and the tails of fishes. In short, there were creatures with the limbs of every species of animals. Add to these fishes, reptiles, serpents, with other wonderful animals, which assumed each others shape and countenance. Of all these were preserved delinations in the Temple of Belus at Babylon.

    Thus we have the written words of Oannes, about mankind and other animals Our ancient ancestors inform us that Oannes came from the star system Sirius. That these superior beings are responsible for the Earth and mankind. When we look at the written words of Oannes and the physical and internal descriptions of men and women, we are quite shocked. Oannes informs us that humans had two wings, two faces; one of a man one of woman. He tells us that these two heads were on one body, and that several of their organs were both male and female! Humans are described as having the bodies of other animals!

The shocking descriptions of human animals, goes along ways in explaining the behavior of humans all over the Earth today. Men and women both shared the same body and their internal organs, which could well explain human conflicts to assume that if males and females were both sharing one body, then the feelings and natural desires of their separate gender, could certaintley have been shared by the human animals. The profound and shocking descriptions of these human animals, could well go along ways in understanding the conflicts between men and women, in our modern world. This reminds me of the old expression "You can't live with them, you can't live without them." When we examine the statement of Oannes, that humans had the body parts of many other animals then we must consider the possibility that the ancient predatory as well as docile behavior of various animals has continued to either haunt or follow mankind to this very day. We are informed that men had the bodies of many animals, this statement would explain why there are vicious unremorseful and predatory humans among us today, who kill for the thrill of killing other human beings. One must assume that humans at this remote time in the deep dark past, must have been part predatory beast such as a wolf or tiger In any case, Oannes describes human animals of every type and countenance. By the same token humans were also part of the gentle species of the wild kingdom. These animals would have been deer, sheep, cows and other rather docile and gentle animals of nature. Mankind therefore did also inherit, the calming qualities of the animals which would be the noble and civilized qualities that the human race is also known for. In short Oannes informs us that there were many humans that had the bodies of every other known animal, therefore we must assume that the predatory as well as the gentle nature of these ancient animals

147

has never left the DNA that courses through our brains. Oannes says that humans at one time possessed the countenance of other animals therefore we must assume that these human animals were hunting and literally eating each other. These human animals had to eat something to survive, so they well could have been eating their fellow human animals. Humans are described as having wings, and this describes the ancient human descriptions of superior type of humans who could fly because they had wings on their backs. These creatures are better known to history as angels. Man has always had a great desire to fly into the sky. Today flying is a very common experience all over the world, and if it were not for mans insatiable need to fly; we would live in a very primitive world. The description that Oannes gives about humans having wings, would certainly account for modern man's great desire to fly. All of the attributes that he describes of the human animals do in fact make a great deal of sense, in light of the behavior of modern man. When we analyze his statements on the fact that there were humans that were both male and female, this would explain why people are having such a problem with homosexuality and it's related issues. Oannes alludes to the fact that there was sodomy involved among the various animals, in fact it would have had to have been sodomy on a grand scale in order to create the many human animals that he talks about. What he describes is not surprising at all when it comes to sodomy. To this very day, there are nations and people all the Earth, who have laws and taboos against sodomy. Oannes is therefore exactly correct when he describes the very nature of sodomy, in those ancient days. We then realize that what Oannes is describing to us is actually a problem that has been with mankind for a very long time. Modern laws against sodomy actually fit right into the ancient words of Oannes, therefore if there were not a problem with this kind of behavior that goes very far into man's ancient past-there would be no such sodomy laws. What has actually happened is that the ancient DNA of our human animal ancestors has never completely left our bodies and especially our "brains," but has followed modern man into the world that we live in today. The ancient human animal DNA in our brains, therefore is the cause of why there are ferocious predatory mass killers, as well as gentle and civilized people; who genuinely care for and help his fellow man. It is because of the ancient DNA that some people were born to be of certain sexual origin, which is causing problems with other people, who condemn these people.

Nostradamus: The American One Dollar Bill And Harry Truman

In the words of the Ancient Egyptians, we find that they are remarkably similar to the discoveries that modern astronomers are making today. All of the ancient sacred Egyptian texts speak of the fact that God was born from nothing more than dark invisible matter, which is what astronomers are now discovering the vast universe is made of. All of this brings to mind The Ancient Trismegistic Literature. This very ancient literature says that mankind will chase out into the height(space) to observe the motions of the heavens. This literature which has turned out to be correct, also says that man will discover the secrets of Isis and the star Sirius. In leaving mankind the ancient Trismegistic Literature, there were also other sources of knowledge that were left for man. For instance the ancient Dogon people inform us that the superior beings from Sirius told them that some day they would return from the stars, and meet the Monitors that they left to watch over man and the Earth. Together they would return to Earth and bring with them, the ancient ancestors of man. They would signal their return to the Earth and the Dogon people, by the appearance of an unknown star in the sky, which will be visible.

In early 1993 a group of scientists discovred an unknown comet in our solar system. This comet in 1992 flew so close to Jupiter the eye star, that it's powerful gravity literally broke the comet into about a dozen large chunks. This previousley unknown comet returned to Jupiter on July 20,1994,

Nothing like this catastrophe has ever been observed before, so close to the Earth! Is this the star that the Dogon people were told would be visible, which would signal the beginning of the return of the superior beings from Sirius? Scientists determined that this comet was broken into about 21 large chunks, therefore when the comet returned during the summer; because it was broken into so many large chunks this celestial body appeared as a star in the sky. The return of this unknown comet returned during the month of July, which is the 90 days of the Dog Days of summer; when Sirius rises in our eastern sky. Coulld this celestial body be the star that the Gods from Sirius spoke of, that would herald their return to Earth in this century?

Other sources of knowledge were left for mankind to inform him of the return of the Masters Of The Waters. Nostradamus is that other source. Nostradamus was born in 1503, and in, it will be 500 years since his birth.

He made his predictions by placing a bowl of water on a tri-pod, and he then looked into the bowl of water. The number 3 which represents the three legs or the pyramid shape, as well as the water identifies the Masters Of The Waters as well as Sirius. Nostradamus said that the United States was the focus of his prophesy. One of the modern signs of that prophesy that ties the United States to the prophesy, is the American one dollar bill. To the ancient Egyptians, the great pyramids and the single divine "Eye" of Amen Ra was representative of their religious worship of Sirius of the great chasms of time; and has manifested itself on the face of the American one dollar bill. The American founding fathers loved and adopted the ancient classic Greek style of government as well as the Greco-Roman style of architecture. The designers of the great seal of the United States, back in the 1700's admired the ageless permanence of the great pyramids and they therefore adopted many of the ideas and the physical symbols of the ancient Egyptians, Greeks, and the Romans. What they did not realize is that they were adopting ideas, signs and physical symbols that are related to the oceans, sea's, fish and ultimately to Oannes and the star system Sirius. The adoption of these ancient symbols has been called tradition. On the American one dollar bill, the ancient pyramid and it's single eye, as well as the ships rigging; are all related to the prophesy of Nostradamus. The Rosetta stone as, the symbol of Egypt and the bridge to the Greek Civilization, is the link to America by way of the pyramid and the single eye on the American one dollar bill. The oceans, sea's, fish, amphibians, sea shells, and the Greco-Roman classic style, as well as the Rosetta stone; links the ancient Egyptians and the Greek civilization to Oannes and Sirius. From here the volcanic stone tablet leads to the Earth's volcanoes and earthquake zones around the world. From here the Rosetta stone links mankind's ancient civilizations and temple sites from all over the Earth, to the invisible electromagnetic pyramid images of the Earth-and the single pyramid image on the planet Mars. From here the electromagnetic pyramid images then lead to mankind's ancient worship of nature and the Earth. From the ancient pagans all over Europe to the indigenous American Indian peoples of the western world, as well as the Island peoples of the entire world. Humans all over the Earth worshipped mother Earth, father sky and sacred mountains. From here the volcanic Rosetta stone points the way to Easter Island and it's volcanic statues, which leads to the terra formed fish, dog, pyramid and bull shaped continents of the Earth. The Rosetta stone also leads to the terra formed continent of Antarctica, which represents the Easter Island statues. From Easter Island the ancient stone takes us to Jupiter, Sirius and Oannes and the secrets of the universe, and the prophesy of the Virgin Of The World.

The Peculiarities Of Peoples Under The Influence Of The Pyramid Images

When we examine the locations of the invisible electromagnetic pyramid images all over the Earth, we discover that there are whole countries and inhabited locales of the world that are completely immersed in the bathing energy of these world wide images. Some of the countries that are almost totally covered by the electromagnetic pyramid images are Japan, Korea, the Philippines, Italy, Germany, Yugoslavia, Austria, Sweden, Finland, Romania, France, Hungary, England, Ireland, Wales, Poland, Czechoslovakia, Norway as well as other areas of the world such as California. Many of these countries and their peoples are known for their certain personality and character traits. For instance the Japanese people are known to be very homogeneous, and they try to shut out as much foreign influence as they can. The Germans are known to be very hard working industrious people. The Finns are known to be a nation of very moody and somber people, who just want to be left alone. The Swedish people are known for their liberal approach to life, and the British are the scions of polite manners. The French people have a reputation for being the stalwarts of aristocracy and elegance, mixed with a touch of arrogance.

When we take another look at many of these countries of western Europe, we find that from this area of the world men have been colonizing the rest of the world for 2,000 years. Investigating these countries further we then understand that it is from these locations that there have been great wars, revolutions and liberations. From some of these countries there emerges a pattern of great ethnic and racial tensions as well. It is from these various regions of the world, that men spread out across the Earth and built the great Western Civilization that we know today.

The Nostradamus Prophesy and the 6th pyramid

We now arrive at the Nostradamus prophesy, after our long and mysterious journey's through mankind's ancient worship of the Sirius religion and it's Gods. As I spoke earlier about Nostradamus, we know that he placed a bowl of water on a tripod and then looked into the water, and was able to speak about the future. The implications of looking into the water and the three legged tripod, are quite indicative of the ties to the superior beings from Sirius and man's ancient beliefs. Nostradamus foretold.the following events in his prophesy.

Quote:

Near the Rhine from the Austrian mountains. Will be born a great man of the people, come too late A man who will defend Poland and Hungary, and whose fate will never be certain Beasts wild with hunger will cross the rivers. The greater part of the battle will be against Hister.

Near the harbor and in two cities will be two scourges the like of which have never been seen. When those of the northern pole are united together In the East will be great fear and dread. One day the two great leaders will be friends. Their great power will be seen to grow. The new land will be at the height of it's power. To the man of blood the number is reported. In the year 1999 and seven months From the sky will come the great king of terror.

The new land was a common term used by Nostradamus, which was identified as America. Nostradamus is talking about America and Russia becoming friends and together they will face an adversary from the East, who will arrive from the "Sky." Currently there are no powers in the future, that can seriously challenge the military power of the United States. It is a well known fact of science that Sirius is in the East, and our ancient ancestors inform us that they will return to Earth from the sky. Sirius is visible from July to September in our eastern sky. Nostradamus says that the great king of terror will arrive in the 7th month of the year 1999. The 7th month is the beginning of the celebration of the birth of America. The number 7 is an ancient sacred number, and it represents the ancient Egyptian pyramids. When we look at the year 1999 and the numbers that are involved in the prophesy, we realize that there are 3 nines in the arrival date. The number 3 represents the Earth and the number 9 represents the Egyptian Great Enneid, or the 9 great Gods of Sirius. The other 9 represents the nine light years that Sirius is from the Earth. The final 9 represents the nine months gestation period of the human fetus. When we examine the year 1999, we then learn that in the following year, we will then be entering the third millennium. Osama Binladen and Terrorism.

The question that we must ask ourselves is, why is America and Russia the focus of the Nostradamus prophesy? The answer to that question may have several aspects, rooted in mankind's past history. Our ancient ancestors and their written artifacts, say that when the superior beings or Gods came to the Earth, they gave mankind a set of laws to live by. Those laws were similar to the laws that we live by today, and the most important one was that man was not supposed to kill other human beings. When we look at the 20th century, we know that between World War I and II, as well as all the other wars of this century the 20th century has been by far the bloodiest century in man's history. When we combine this with the fact that mankind used the sacred white light of the stars and the Gods, to kill other human beings then we can begin to understand the ancient laws that were bequeathed to man. Ancient written artifacts proclaim that when events on the Earth get out of hand, The Gods will return to our planet. To better understand when events get out of control on the Earth, and the relationship to the sacred numbers 2, 3, 5, 7 and 9; we now must examine the number 6. Our ancient as well as modern man, all are aware that the three numbers of 666 are generally considered evil. The 3 numbers of six are representative of the Earth, and they represent the evil that man does to his fellow man. When we peer even closer at the number 6, we find that there is even more startling events surrounding the number 6 here on the Earth! To understand the greatest wars and evils of the 20th century, we must return to the six electromagnetic pyramid images that are located all over the Earth. When we take a closer and more detailed look at the number one pyramid image, and the third electromagnetic pyramid image, which intersect each other in Western Europe suddenly a most startling event has taken place! There is now the unexpected appearance of a 6th pyramid image on the surface of the Earth! This 6th pyramid image is located in many countries of western Europe! Everyone is aware that the number 6 is associated with evil. The sudden emergence of this 6th pyramid image in Europe incredibly identifies the "Axis" powers of World War II, who were responsible for the greatest carnage of the 20th century! Astonishingly, this 6th pyramid also correctly identifies the beginning location of World War I as Sarajevo, or the former Yugoslavia! World War I began when in June of 1914, Austrian Crown Prince Francis Ferdinand was assasainated by a Serbian terrorist in Sarajevo, the former Yugoslavia:

The unprecedented killing of human beings, had now begun to define the 20th century. After the carnage of World War I the unparalleled horrors of World War II began with a man named Adolph Hitler, which Nostradamus speaks of.

The 6th pyramid, which is evil in nature correctly identifies the various governmental leaders and countries that were responsible for starting both World War I and World War II! The three boundaries of the large pyramid image correctly identifies the instigators of the wars in Europe as well as the Pacific! The base of the 6th pyramid image faces all the way across Europe and Russia, and points right at Japan! Because of the horrible events of the 20th century, mankind may be called to answer for his evil. When we take another look at many of these countries of Europe, we find that from this area of the world men have been spreading out to other areas of the world for 2,000 years. Realizing that there are other ethnic hatred's in the world, we then realize that from this area there exists great ethnic and racial hatred's that have been taking place among peoples for thousands of years! These very animosities are continuing to this very day in Bosnia Hercegovina.

Ancient religious sacred texts of the Sirius religion, inform us that man is supposed to follow the laws that the superior beings from Sirius gave to mankind. These laws were intended to help mankind to live in peace with his fellow man. They were supposed to show men how to live in understanding of one another. In short, the laws that were given to the people of the Earth were intended to help us to be civilized. The ancient Egyptians speak of the sacred white light of the stars and the Gods. When mankind learned how to unleash the power and the sacred white light of the Gods, at the Trinity bomb site, events then as well as before were set into motion; for the return of the Gods from Sirius according to our ancient ancestors. Because ancient written artifacts and peoples from the distant past, all proclaim that modern man is now in the fifth civilization, it is because of the behavior of man that they inform us that the Gods will return to Earth with the ancestors of man!

Apollodorus Quote: "Sky was the first, who ruled over the world"

The Eternal Ancient Sirius Religion: In the ancient Egyptian worship of Sirius the Egyptians were completely obsessed with death and the after world of which every human being is destined to experience. All along the 6 pyramid images of the Earth, death and its consequences have been the single greatest events that have taken place on the world wide pyramid images. The universally recognized Egyptian pyramid and the death events on the 6 pyramid images throughout history all confirm the ancient Egyptian obsession with death and the underworld as the main theme of the Sirius religion.

Hermes: Son Of Zeus And Maia

Our ancient ancestors reveal to us that they left written artifacts and messages, for mankind to decipher. Part of those messages are the ancient Gods and what they stood for.

Hermes: (Greek mythology) Son of Zeus and Maia, messenger of the Gods and the God of wealth, luck, and roads and conductor of souls to Hades. He is identified with the Roman Mercury.

Hermes Trismegistus: The Greek name for the Egyptian God of wisdom, Tho th, identified with the Greek Gods.

Hermes. The Neoplatonists attributed the Hermetic books, works containing the collected occult knowledge of ancient Egypt to him.

Maia: (Greek mythology) The daughter of Atlas and the mother of Hermes. She was the eldest of the Pleiades.

Pleiades: The seven brightest stars in the constellation Taurus. (Greek mythology) The seven daughters (Alcyone, Celaeno, Electra, Maia, Merope, Asterope, Taygete) of Atlas who were changed into stars by Zeus

Taurus: A northern constellation. The second sign of the zodiac, represented as a bull.

Sirius: Sirius is a binary star system that is 9 light years from the Earth. Sirius A is a blue white star that is larger and brighter than our own yellow sun. Sirius B is a white dwarf which consists of electron degenerate material. It is extremely dense and heavy. It takes fifty years for Sirius B to orbit Sirius A.

Southern Cross Constellation: A cross shaped constellation in the southern hemisphere with a bright star at each extremity. It is the symbol of water.

Condensed list of Legends, Historical Events And Tragedies, That Have Occurred On The 5 Electro magnetic Pyramid Images All Over The Earth; With An Emphasis On The United States.

MARS: Four giant volcanoes and their location on this planet, form a pyramid image. Mars is the symbol and God of war and destruction. These are the events that Pyramid #   1—have taken place on Earth!

Location: Giza Egypt to the Aztec empire of Mexico City, then to the Inca Empire at Machu Pichu

Texas:

1.   Mysterious Marfa light of Texas observed for centuries.
2.   1961 University of Texas mass murder, kills many students from a campus tower.
3.   The 1963 assassination of President John F. Kennedy, in Dallas Texas.
4.   In the mid 1980's there were a number of tragic plane crashes that are attributed to severe weather and wind shear.
5.   1991, a Texas man drives his truck into a restaurant cafeteria and kills many customers inside. This happened in Killeen Texas
6.   In 1992, there was much flooding in the state, that lasted long past the normal rainfall
7.   In 1993, the Branch Davidian Sect of Waco Texas, burned to the ground, with the loss of many children. Waco (we ain't coming out)
8.   There have been cattle mutilations in this state.
9.   Texas is known for very severe thunder storms.
10. Texas is known as being in the Tornado Alley belt

11. North west Texas is in the Bible belt and Texas City ship explosions that killed 500
12. The Alamo
13. Columbia Space Shuttle Disaster

Missouri:

1. President Harry Truman was from Missouri, he was the first human being to order the use of atomic weapons.
2. The mysterious New Madrid earthquake fault zone is located in this state.
3. In 1811-12 the New Madrid fault created the largest eart hquakes in this country. There were 3 gigantic quakes
4. Missouri is known for having very powerful lightning storms.
5. There have been cattle mutilations in this state.
6. Missouri is also located in Tornado Alley.
7. In the summer of 1993, Missouri experienced the largest flooding that has ever been seen in this state.
8. 1981 Hotel Collapse in Kansas City 114 dead

Oklahoma:

1. Oklahoma is known for it's severe lightning storms.
2. There have been cattle mutilations in this state.
3. In June of 1993, an Oklahoma man killed all of his family members in a mass slaying.
4. Oklahoma is located in the Tornado Alley
5. Oklahoma is part of the Bible belt
6. Timothy McVeigh and the Oklahoma City bombing
7. Tulsa Oklahoma race riots,

Tennessee:

    1. Dr. Martin Luther King was assassinated in Memphis Tennessee.

Ohio:

    2. President James Garfield was ass assinated, he was from Ohio.

    3. President William McKinley was assassinated, he was from Ohio.

    4. In the 1930's, the Cleveland mass murderer was never caught. He was known as the torso murderer.

    5. In 1989, a Cauldwell Ohio man was involved in the mass murder of 5 people.

Illinois:

    1. President Abraham Lincoln was assassinated, he was from Illinois.

    2. In 1871, the great Chicago fire burned down 17,000 buildings. It burned for 27 hours (March 8th 1925 695 dead from tri-state tornado striking Indiana, Ohio, Illinois. 243

    3. Legend has it that Mrs. O L eary's cow kicked over the kerosene lamp, that started the great Chicago fire.

    4. Chicago for 100 hundred years has been the largest cow and bull meat processing center in the world. (the cow and bull are sacred animals in the Sirius religion)

    5. Enrico Fermi and his colleagues at the University of Chicago, were the first scientists to achieve a controlled nuclear reaction.

6. In 1961, the mass murderer Richard Speck was apprehended by the police in Chicago

7. In the mid 1970's John Gacy was convicted of the mass murder of many young males.

8. In 1979, the worst Jet airliner crash occurred at O'Hare International airport.

9. Atomic research at the University of Chicago and the Manhattan Project

10. In 1992, an unknown mass murderer, killed 7 young workers at a suburban Chicago restaurant.

11. St. Valentines Day Massacre.

12. June-July of 1993, Illinois as well as surrounding states, suffered catastrophic flooding.

13. Dr. Clyde Tombaugh discovered the planet Pluto, which is the 9th planet. He was from Illinois.

14. The detection of one of the 5 electromagnetic pyramid images by scientists

Wisconsin:

1. Like Chicago Illinois, Wisconsin is known as the cow state, and is famous for it's cheese.

2. Fond Du Lac Wisconsin, known as "The Miracle Mile."

3. This area is known as the location of more million dollar lottery winners, than any place in the nation.

4. A young teacher and his future wife won the largest single lottery jackpot in U.S. history. 110 million dollars

5. In 1992 Jeffrey Dahmer was exposed as a cannibalistic mass murderer.

Michigan Great Lakes region and North East:

1. The Great Lakes region is famous for the greatest number of mysterious disappearances of ships, planes, and people. Gordon Lightfoot sang a song called "The Edmund Fitzgerald."
2. A mass murder occurred at a U.S. Postal Office In 1992 in Michigan.
3. New York, the World Trade Center attacks and the Pentagon
4. Egypt and Swiss Air plane crashes, 460 dead
5. John F. Kennedy Jr. plane crash, earthquakes in the North East United States
6. Nor Easter in film "The Perfect Storm"
7. American Revolutionary War "The Patriot"

Canada:

1. This is the location of the Lake Champlain sea monster, who is the cousin of the Loch Ness monster. It's name is "Champ."
2. Quebec Canada, this is the location of so much strife. The
3. French speaking people of this region are speaking about seceding from Canada.
4. Bigfoot sightings

Labrador:

1. In 1941 a whole flight of American war planes mysteriously crashed in Labrador, while on their way to England.

2. In 1985, over 200 U.S. servicemen were killed when their plane crashed at Gander New foundland, while on their way home from Egypt.

North Atlantic Ocean

1. The Titanic ocean liner, while on it's maiden voyage from England, sinks in the North Atlantic Ocean; after it struck an iceberg. The liner is carrying the ancient Egyptian mummy princes Taheb.
2. WW II Nazi Wolf packs attacks on Liberty Ships

Scotland:

3. Scotland is known as the location of the famous Loch Ness sea monster called Nessie
4. In the mid 1980's terrorists blew up a 747 jet airliner over Lockerbie Scotland.
5. The Film "Braveheart"

Wales:

1. The detection of one of the 6 electromagnetic pyramid images of the Earth, by a group. of scientists.

Ireland:

2. Ireland is the location of a civil insurrection that has been going on for over 25 years. The Irish Republican Army has been bombing targets in Northern Ireland as well as downtown London.

England:

3. England is the location of many mysterious and unidentified peoples such as the Druids

4. Stonehenge is located here.

5. England is the location of many ancient Roman temples.

6. Like the great fire of Chicago, London also had it's own great fire of London.

7. London as well as England was the location of the great Bubonic plague.

8. London was the location of the infamous Jack The Ripper.

9. For over 100 years, England has been the location of the greatest concentration of spontaneous human combustion victims.

10. For centuries, England has been famous for it's many castle hauntings and ghost sightings.

11. England is the country where the British Terrier dog originated from. This dog is the most constant barking dog in the world. Our ancient ancestors inform us that the dog was given to man, to protect man from unseen evil spirits; by barking at the spirits and scaring them away from man. In England, this dog has only been doing what it was in tended to do, because of all the evil spirits there.

12. In June of 1993, London has been the sight of a homosexual mass murderer.

13. Spring Heeled Jack the mysterious disappearing creature and the devils feet of Britain

14. England has a very violent past history.

The 6th evil pyramid location:

1. The evil 6th pyramid location, correctly identifies Yugoslavia as the location of the beginning of World War I, at the dawn of the

20th century. It also correctly identifies Yugoslavia as the killing grounds at the end of the 20th century.

2. The 6th evil pyramid that is located in Europe, correctly identifies the Axis Powers of World War II. These belligerent countries were Germany, Italy, Japan and Austria

3. The base of the 6th pyramid faces Japan.

4. Enrico Fermi was born in Italy and studied nuclear physics in Italy. He was connected with the atomic bomb

5. Albert Einstein was born in Ulm Germany. He was the famous nuclear theorist who sent President Roosevelt a letter outlining the possibility of building a nuclear weapon.

6. Roosevelt then authorized the top secret Manhattan Project, to develop the atomic bomb.

7. Some of mankind's greatest disasters have taken place in Europe

Nostradamus And His Prophesies Which All. Occured On The 5 Electromagnetic Pyramid Images Of The Earth

Pyramid #1

1. Nostradamus correctly prophesied the death of Henry II of France in 1559.

2. Nostradamus correctly prophesied the capture of Louis the 16th, and Marie Antoinette of France

3. Nostradamus correctly prophesied the rise of Napoleon of France

4. Nostradamus correctly prophesied the rise and fall of Charles the I of England in 1645

5. Nostradamus correctly prophesied the great plague of London 1665

6. Nostradamus correctly prophesied the great fire of London of 1666

7. Nostradamus correctly prophesied the medical contributions of Dr. Louis Pasteur, and the pasteurization of cow's milk

8. Nostradamus correctly prophesied the abdication of Edward the 8th of England, to marry his fiance

9. Nostradamus correctly prophesied the emergence of General Franco, and the civil war in Spain in 1936

10. Nostradamus correctly prophesied the rise and fall of Adolph Hitler during World War II

11. Nostradamus prophesied the arrival of the great king of terror, who will come from the sky from the east in 1999

12. Nostradamus correctly prophesied the 2 atomic bombs of WW II

Pyramid #5

Location: Washington state to California peninsula, then to the eastern plains of Colorado.

Washington:

1. 1945, Hanford nuclear processing plant, is the world's first nuclear plant.
2. In 1947, 9 flying disks were sighted over Mt. Rainier. The modern term flying saucer was coined.
3. Washington state is the location of the mass murderer, Ted Bundy.
4. This is the location of the Green River mass murderer.
5. Mt. St. Helens erupted on May 18, 1980.
6. Bigfoot sightings
7. The detection of one of the electromagnetic pyramid images

California:

1. The great San Francisco earthquake of the 19th century.
2. In 1968, Robert F. Kennedy was assassinated at the Ambassador Hotel in Los Angles.
3. In the early 1970's, in Yuba City, a mass murderer killed many migrant workers in California.
4. On March 13, 1974 an airliner crashed in Bishop California, no.reason could ever be found by the FAA why the mysterious plane crash took place.
5. In 1987, a man armed with automatic weapons entered a McDonald's restaurant and killed many people. The restaurant was later torn down.

6. In the early 1980's two men in eastern California are involved in the mass slaying's of their neighbors. One man swallows cyanide, the other man is captured in Canada.

7. In 1990, Richard Ramirez is captured and convicted of mass murder in southern Californis.

8. In 1991, mass murderers were apprehended at a Sacramento mall.

9. In 1991, the greatest fire to strike the U.S. occurred in southern California.

10. In 1992, Dorothy Fuente was indicted in the mass slaying of elderly people for their social security checks.

11. In 1991, there was a mass murder at a U.S.Postal office in California.

12. In 1992, the 7 year drought in California came to a halt when torrential rains and flooding hit the state.

13. On July 2, 1993 a gunman in San Francisco walked into a high rise building and killed 9 people including himself.

14. The 5th pyramid image travels through southern California as well as Nevada where nuclear weapons are tested.

15. Bigfoot sightings

16. Charles Manson family and mass murders.

17. Large fires and riots

18. Alaska Air Lines MD 80 crash

Mexican border:

1. In the late 1980's a cult of devil worshippers, killed many victims for the purpose of sacrificial offerings

New Mexico and Arizona:

2. In 1942, Los Alamos was the location of the Manhattan atomic bomb program.
3. The first atomic bomb was exploded at the Trinity site near Socorro, in 1945
4. Socorro New Mexico is the location of the most celebrated UFO incident.
5. The White Sands missle range is the location where this country first tested missles for the dawn of the space age.
6. Wyatt Earp and the O.K. Corral and Tombstone
7. Tombstone and Boot Hill
8. Geronimo and Chochise Indian wars
9. The Taos area is the site of a mysterious "humming" sound that nobody can figure out..
10. Albuquerque, site of the mysterious "humming" noise
11. Billy the kid and Lincoln county wars.

Colorado:

1. The sand dunes of Alamosa Colorado are located here. They have black magnetite that has a positive magnet is attraction to the pyramid image in Colorado.
2. Colorado is part of Tornado Alley, and it has the third most tornados in the United States.
3. In 1991, a B1B Air Force bomber crashed north of La Junta Colorado.
4. In 1992, while on approach to the airport in Colorado Springs; a 737 jetliner mysteriously crashed. The FAA could never find a reason for the accident.

5. In 1989, the most severe hail storm to strike the United States, struck Denver and surrounding areas. The storm caused 250 million dollars in damage.

6. In Manitou Springs and outlying areas of Colorado, there is a 13 degrees magnetic disturbance.

7. Near Greely Colorado, there are large cow meat processing facilities.

8. The single largest lottery winner in Colorado history, bought her winning ticket in Boulder Colorado; on the pyramid image.

9. In this state there have been cattle mutilations, and recentley they have begun again.

10. In Colorado large computer systems go "down" for no apparent reasons.

11. The computers at NORAD have been having trouble operating to design specifications for years.

12. Diamonds which come from volcanoes, are located between Colorado and the Laramie Wyoming border.

13. The traffic computers of Colorado Springs have never operated to design specifications, consequently many citizens complain about the timing of the lights.

14. Bigfoot sightings

15. Columbine school massacre

16. Colorado Springs 737 plane crash

17. Sand Creek Massacre

18. June 8, 2002 Massive Colorato Forest Fire

Wyoming:

1. In 1979, an amateur photographer was able to take the picture of a 1,000 ton meteor as it streaked over Wyoming.
2. This state has had cattle mutilations.
3. Bigfoot sightings
4. Mega volcano called the "Beast" in Yellowstone and 1988 fire
5. Custers 7th Cavalry and Battle of the Little Big Horn.
6. Indian Wars, Sitting Bull, Crazy Horse, Red Cloud
7. Western Law Breakers including Butch Cassidy and the Sundance kid.

Utah:

1. This state is the location of the Great Salt Lake. Salt and natron was sacred to our ancient ancestors, and was used in the process of mummification, it was sacred in ancient times. The phrase "A man worth his salt," was coined by ancient man.
2. Bigfoot sightings
3. Mormon violence in the past history

Montana:

1. Flathead Lake, Loch Ness sea monster sightings
2. The bloody Bozeman trail

Pyramid # 2

Location: Kamchatka Peninsula to Tambora Indonesia, then to Easter Island.

3. The testing of nuclear weapons was done in the Marshall Islands, which is right in the middle of this gigantic pyramid image. The ancient Egyptians buried the Pharaoh in the middle of the pyramid.

4. Hiroshima is the location of first atomic bomb droped on man.

5. On December 7, 1941 the Japanese attacked Pearl Harbor, thereby forcing the United States to enter World War II. Nagasaki Japan is the location of the second atomic bomb to be dropped on mankind.

6. The Korean war was fought off the coast of Japan.

7. The Viet Nam war was fought off the coast of Japan.

8. In 1985 the Japanese KAL jet liner with over 250 was shot down by Russian Migs over Kamchatka.

9. The detection of one of the 5 electromagnetic pyramid images in Queensland Australia

10. Cambodia, the Killing Fields

Pyramid # 3

Location: Italy to India then to the Aleutian Islands.

1. Tunguska Siberia was the sight of the giant meteor incident

2. The Himalayas are the site of the Abominable Snowman

3. The 1980's Afghanistan war with the Russian army, was a central reason that the Communists went economically broke, thereby causing the fall of Communism; and the fullfilment of the Nostradamus prophesy.

4. This is the route of Alexander the Great to India 500 B.C.
5. This is the route of modern day terrorist states Iraq, Iran, Saudi Arabia, Afghanistan, Pakistan etc.
6. Route of Genghis Khan to Europe
7. Route of the Vikings to Europe

Pyramid # 4

Location: Equatorial Guinea Africa to Tristan da Cuna to Mascrene Islands.

1. South Africa, location of apartheid and great injustice.
2. Boer and Zulu wars to pyramid #4

6th pyramid:

1. This pyramid image is located in Germany where the NATO and Warsaw Pact forces were locked in a Cold War, since the end of WW II. The Berlin wall was a symbol of that confrontation. Eventually good triumphed over evil
3. The reemergence of neo-Nazism in Europe again.
4. Mankinds greates disasters.

Mankind's ancient expressions that have their roots in the Sirius Gods and their religion

1. Holy cow
2. Take the bull by the horn
3. Dog days of summer.
4. Thank my lucky star
5. What goes around comes around
6. You can't see the forest for the trees
7. You can lead a horse to water but you can't make him drink

8. Holy mackerel
9. Sprout wings and fly
10. Go to Timbucktu (Mali Africa)
11. Wave a magic wand
12. My guardian angel
13. Top dog
14. King of the mountain
15. A mother's intuition
16. You animal!
17. It's written in the stars
18. You can't live with them You can't live without them
19. I was born to?
20. That's fate
21. I have a calling
22. Work his magic
23. It's not in the cards
24. We are all pawns
25. It's a sacred cow
26. It's a cash cow
27. You missed the boat
28. The writings on the wall
29. Dead in the water
30. The face of the Earth
31. One with nature
32. The man in the moon
33. We are all in the same boat
34. History repeats itself
35. That's ancient history
36. In hot water

37. A God send
38. Free spirit
39. Stay for a spell
40. Clam up
41. Old wives tale
42. Father time
43. Wisdom of old age
44. Under a spell
45. Your a horses ass
46. God fearing
47. Working my tail off
48. Spook them out
49. Animal magnetism
50. Three coins in the fountain
51. The well of human kindness
52. A man worth his salt
53. Eternal triangle
54. Art is a mirror of life
55. By word of mouth
56. Star light star bright
57. I wish I may I wish I might
58. Fight like cats and dogs
59. Come back to haunt you
60. It's raining cats and dogs
61. The fountain of youth
62. The ring of fire
63. Lady luck (The luck of Ani, Egyptian Godess)
64. The light at the end of the tunnel
65. Dog eat dog world

66. It's a jungle out there
67. It's in the blood lines
68. Take her or him under the wing
69. The sky is falling
70. Twist of fate
71. He or she is a pillar of society
72. Life in a fish bowl
73. Keep my fingers crossed
74. Your ship has come in
75. Mother Earth
76. That's water under the bridge
77. Count you're blessings
78. Leave sleeping dogs lie
79. Till hell freezes over
80. Good luck to you
81. God speed
82. Don't cry over spilt milk
83. Blue blood
84. Father sky
85. The stork brought the baby
86. It's clear sailing
87. Blue sky (investment)
88. The grass is greener on the other side
89. That's fishy
90. Fight tooth and nail
91. Things get out of hand
92. Seventh heaven
93. Bad moon rising
94. Golden calf

95. Nine lives
96. Sacred mountain
97. Knock on wood (sacred tree)
98. More than meets the eye
99. The cow jumped over the moon
100. My good luck
101. My bad luck
102. Hocus pocus
103. Come hell or high water
104. Dog fight
105. Dog days moon

Pyramid #1

Location: Aztec empire of Mexico city to Inca empire at Machu Pichu, to the three pyramids at Giza Egypt, and then back to the Aztec empire.

1. This electromagnetic pyramid image and it's location in the United States, correctly identifies the Northern and Southern states; that fought against each other in the Civil War of 1860.
2. This electromagnetic pyramid image correctly identifies the Northern states who were against slavery, as well as President Abraham Lincoln.
3. This electromagnetic pyramid image correctly identifies the Southern states who were for slavery.
4. The location of Tornada Alley.

Sirius And The Earth: Informal Facts And Thoughts

The temperature of the universe is exactly 2.735 above absolute zero. These numbers are the same as the ancient sacred numbers of our ancestors. Sirius is 9 light years from the Earth. Our solar system has 9 planets revolving around the sun. The great Egyptian Enead has 9 great Gods, and the gestation period of the human fetus is 9 months in a complete water environment. The modern term flying saucer was coined when a pilot spotted 9 flying disks over Mt. Rainier. If man does not drink water in 3 days he will die, and the number 3 represents the Earth. The number 5 represents the planet Jupiter and there are 5 pyramid images on the Earth, with the 6th one being evil. Our ancient ancestors inform us that we are in the 5th civilization. The human body and all of it's fluids are salty as well as the flesh. Every human being and his body is naturally receptive to being embalmed or otherwise mummified. We are all humans who are walking around already half mummified! We are amazed when we see the work of taxidermists, and the animals that they preserve. Humans and all life on this planet are no different, in that our bodies are all programed to be mummified; therefore we can go to the funeral home and see our dead relatives who look like they are sleeping. This single fact confirms the ancient Egyptian Sirius religion. In the Nostradamus prophesy, water in a bowl was used by this mysterious man to read the future. When we look at the ancient pyramid, we can see that the Egyptian temple has four sides and the base, which totals 5. Five is the sacred number that represents Jupiter. Jupiters moon Io is completely covered with volcanoes and lakes of lava fire. The ancient Dogon people are the modern day human advocates of the superior beings of Sirius. Their entire lives revolve around the Sirius religion. Originally they lived near the 5th cataract of the Nile river. Experts tell us that the world's population is 6 billion people. The 6 billion people of the Earth, may well indicative of the 5th civilization that our ancient ancestors tell us that we are in. In the Nostradamus prophesy, he tells us that the "number" is reported and he is called the man of blood. Currently the Earth is straining to support a vast Earth population, and mankind is under assault by a pandemic biological agent called AIDS that threatens every human being on this planet. It is chilling that Nostradamus made references to a number of powerful events that are happening today. In his prophesy he says that the two leaders will be friends and this prophesy has already come true. The United States under Bill Clinton was the benefactor of the new Russina government. Throughout this entire book I have spoken about our ancient ancestors, which are the people who lived before Christ or B.C.. Bill Clinton or B.C. certaintly has the appropriate initials and his role in befriending the Russian people, identifies him in the Nostradamus prophesy. During Bill Clinton's presidency, the Midwest has suffered the greatest flooding that this country has ever seen. The flooding actually created a 6th Great Lake in the middle of the United States. This brings to mind the fact that there are 5 Great Lakes in the northern part of this country, and the significance of this number is not lost to us. When we look at the volcanoes all over the Earth, the facts tell us that 80% of the world's volcanoes are located under water, thus confirming again that the volcanic Rosetta stone is associated with The Masters Of The Waters. Briefly I mentioned that our human bodies are naturally receptive to embalming or Egyptian mummification. When we look at the human body, we realize that all of the various liquids in our bodies contain salts, not unlike that of the Earth. The human body must have certain amounts of salt otherwise the human body and the brain cannot function. When we consider the ancient Sirius religion we know that the Egyptians used the sacred natron salt of the Nile river to embalm the dead Pharaoh. They also used other spices and natural elements as well. One of the reasons that the ancient Pharaoh's were able to last so long, in their embalmed state; is that the human being is pre-dispositioned to being embalmed or mummified! This is because the flesh of a human being is salty, and in reality; every human being is already half way to being mummified. Most people who die are embalmed and this is really no different from mummification, except that the ancient Pharaoh's were intended to last much longer; than traditional modern embalming. The difference is that humans today are not wrapped in linen bandages. When we look at what parts of the body that the ancient Egyptians preserved, we learn that they mummified the heart, lungs, liver, and intestines. However they did not preserve the brain, why not? These other organs were either placed in canopic jars or in canopic sacks. In the ancient Sirius religion these various body parts were sacred therefore they were preserved for the next life. For instance every human being has 50 feet of intestines, and this is the exact number of 50 years that it takes Sirius B to orbit around Sirius A. In addition the orbital configuration of Sirius A and B is an ellipse, that is the exact shape of the hat that the Pharaoh wore It was shaped like a bowling pin, and this shape is the exact design of the orbital shape that Sirius A and B form; when their orbits are traced on a piece of paper! When we look at the God Bes, we then notice that he is described as a dwarf being and we also realize that Sirius B is also known to scientists and our ancient ancestors, as a dwarf star. When we study the Bible of the Sirius religion, we learn that the preserving of the various organs, had a definite reason that served various purposes in Amen or Amenta, the Egyptian heaven. On the other hand, the discarding of the brain was a quite routine matter that has never been understood. The ancient Egyptians knew that the human brain was responsibie for thought and other thinking processes. If they did indeed build the great pyramids, which would have involved the great use of many people and their brains from the planning and engineering, to the actual building-why then did they discard the royal brains of the Pharaoh—whose own idea it was to build the pyramids? Throwing away the brain it seems would have been the ultimate insult and sacrilege, to the very human organ that was responsible for the thinking processes that were responsible for the construction of these Sirius religion monuments! The answer lies in the fact that the Sirius religion was a secret and hidden recognition of the Gods from Sirius. Only the secret religious initiates and the Pharaoh, were privy to the entire secrets of who the Gods were and where they came from! This is a common problem that has plagued mankind for a long time. The leaders at the top and in positions of power, have always kept secrets from the people that they rule; in the self deluding belief that what they don't know won't hurt them. The bottom line is that the Sirius religion was a hidden religion that was called the "Hermetic" or sealed from the eyes of others. Hermetically sealed from the general ancient populations, is exactly how the rulers wanted their religion to stay. People of course knew about the Gods and all of the general information, but they did not know the real important details. The ancient Trismegistic Literature is described as being of hidden ancient knowledge, which is exactly what it was. Therefore when we talk of the brain and it's thinking and cognizant abilities, the brain actually posed a threat to the Pharaoh and the religious worship of Sirius! If some intelligent common person were to figure out who the Gods were, then the power base of the rulers would have been challenged. Because of this the brain was hooked with a piece of metal hooks, and pulled out through the nose and thrown away. The cognizant and reasoning abilities of the human brain therefore was feared, because mankind might figure out who the Gods were and then possibly challenge them just as the Giants challenged the Olympian Gods for control of the universe in the mythology of the ancient Greeks! Because mankind has now learned how to unleash the power of nuclear weapons, he may be in a position to challenge the Gods much like the Giants did with the Greek Gods who are actually the same Gods from Sirius. When we look at the ancient Greek mythology, we learn that the Giants were Alcyoneus, Gaia or mother Earth as well as others who challenged the Gods for supremacy of the universe. Gaia as we all know is mother Earth and the protector of mankind and all life on the planet, therefore mankind would be a challenge to the Gods because through the cognizant abilities of the human brain which the ancient Egyptians threw away and feared-men were able to figure out how to harness the sacred white light of the stars, which is what the Gods are made of according to the Bible of the Sirius religion. When we examine these thoughts, we realize that this may be the reason for the Nostradamus prophesy, because mankind is now learning too much about the universe and what the Gods are composed of; and where they come from. In the ancient Trismegistic Literature and Virgin of the World, the unknown author informs us that mankind will in fact discover the greatest secrets about Sirius and the Gods. Those secrets are now being forth coming in the new secrets that astronomers are discovering about the dark invisible matter in the universe, as well as the new theory that matter can be born from nothing. In the ancient Bible of the Sirius religion of The Egyptian Book Of The Dead, scientific statements about the birth of God and what he is made of, corresponds to the astronomical discoveries that are being made today, as opposed to the Gods of other religions who say that God just exists and they cannot offer any qualifying proof that their particular God does exist. The ancient Egyptians and their Religious worship of Sirius, do in fact offer that qualifying proof. When we look at mankind and who has developed the weapons of war, there is no question that men and not women have been the developers of killing weapons. Earlier I mentioned that women were held in high esteem by the Sons of the Gods. The beautiful women of Earth bore the children of these superior beings. Women have never been responsible for the development of weapons of war or participated in them, to the extent that men have. One of the greatest laws of the Gods, forbid humans to kill other humans therefore women are not considered to be responsible for the misbehavior of mankind. We then remember that the location of the 6th evil pyramid image is located in Europe. It is from this area that most of the world's major historical events, both good and bad have occurred. These events like the rise of the Christian church, the dark ages, the Reformation, the Renaissance, the French revolution, World War I and II, the Holocaust and other events are still affecting modern man. It is from the 6th pyramid image that the greatest killing of human beings have taken place. Our ancient ancestors inform us that our Earth is considered unclean Earth and there are legends that tell us that the Earth was created in 6 days which would confirm that mankind has been responsible for great atrocities to his fellow man. Investigating the Nostradamus prophesy, we are all aware that America is the focus of this event. The dawn of the space age has been pivotal in the discovey clues to the Sirius religion, as well as the dark invisible matter of the universe. America has a long and rich history of naming their space boosters after the ancient Gods of Greek mythology. Space rockets were named the Jupiter, Thor, Titan Nike, Saturn, as well as other ancient names. Spacecraft and space probes were named Mercury, Gemini, Apollo, Ulysses as well as other ancient names. This confirms our ties to the ancient Gods, and our continueing ties to them, because over the centuries many new astronomical discoveries have been named for the mythological Gods or their consorts. Looking at America we know that Americans are some of the most fair, generous, honest, and hard working people on the Earth. However this country is also known as the most violent society in the world, with it's great amount of guns that are available to the public. When we remember that there have been many terrible and tragic events on the pyramid images of the world, we then see that in America; there are two invisible electromagnetic pyramid images in the country. This is the explanation for the great amount of violent crime in the country, as well as other factors. When we look closer at the 5th invisible pyramid image in the United States, whose apex is the whole state of Colorado we discover that the longitude line of the Earth that travels through the eastern plains of Colorado-has a very definite connection to the ancient world of Greece. The invisible longitude line travels across the Earth and intersects the ancient Greek country. In addition, the 6th evil pyramid image and it's apex, also point right at the same longitude line. If we look at the globe of the Earth and the position of the 5th pyramid image in the United States, if we could slowly move the pyramid image across the world; and followed the exact same longitude line from Colorado and went east the 5th pyramid image would fit into place at the apex of the number 1 pyramid image that points right at the three great pyramids at Giza Egypt! These facts are all quite ironic just as there are many ironies in the whole Sirius religion. For instance there is a famous picture of President John F. Kennedy shaking hands with President Bill Clinton. This happened 3 decades ago at Boy's Nation. Kennedy was the man who was instrumental in the landing of men on an other world, which was the start of the fulfilment of the Trismegistic Literature and it's prophesy of man's venture into the universe. It was ironic that Bill Clinton, the man who is spoken of in the Nostradamus prophesy as being the friend of the Russian leader also shook hands with the president who was associated with the ancient prophesy of the Trismegistic Literature!

In the book that I have written. I believe that I have uncovered the key to un derstanding mankind's ancient origins, history and the understanding of all of his ancient mysteries; into one comprehensive explanation. That subsequent explanation is rooted in the ancient Egyptian pyramid, Rosetta stone and the ancient Sirius religion. consequently this involves the star system Sirius, and the superior beings or Gods; that ancient written artifacts reveal are the beings who are responsible for mankind.

All along the America's, there exists the ancient temple sites of the Aztecs, Toltec's, Olmec's, Mayas and the Incas. They built great stone pyramids just as the ancient Egyptians did. The ancient neolithic Anasazi as well as other Indian cultures built great stone buildings at Mesa Verde. Chaco Canyon as well as numerous other locations in the western United States. All of these ancient cultures left written records of the existence of the Gods from the stars. All of these mysterious civilizations were located on the number 1 and number 5 pyramid images in the America's. Every one of them mysteriously disappeared for reasons unknown, to modern day anthropologists. There is no doubt that many strange and terrible events have taken place on the invisible electromagnetic pyramid images of the Earth. The star system Sirius and the universal pyramid image, are the keys to understanding the true history of mankind and the universe.

My Green Mountain Falls Home And The Pyramid

There are 3 large granite boulders (igneous or volcanic in origin) that surround my home, and when one looks at these megaliths, they form the shape of a large Egyptian pyramid image. My home sits right in the middle of the pyramid image, that these 3 megaliths form. The apex or top of the pyramid image is a large gray-green granite boulder that is shaped like the head of an amphibian. This amphibian shaped megalith faces east, right at the star system Sirius. The lower part of the jaw or mouth of the amphibian shaped boulder, has broken away to form the shape of an open gaping mouth. This effect has resulted in the shape of a perfect pyramid shaped upper mouth!

When we look back at the Fragments of Berossus, which is the description of Oannes, the superior beings from Sirius who are responsible for mankind we remember the last couple of sentences of the words of Berossus.

Quote:

When the sun set, it was the custom of this Being Oannes, to plunge again into the sea, and abide all night in the deep for he was amphibious.

Source: Fragments Of Berossus From Alexander Polyhistor Of The Cosmogony And Causes Of The Deluge

When we look at the location of the 3 megalithic granite boulders, that form a pyramid image we learn that my home is only 3 blocks from the water falls that this town was named for. In reality the water falls near my home is actually called a cataract, which means a body of water falling over a ledge. In ancient Egypt, there are 5 recognized cataracts on the Nile river. It is ironic that the water falls is 3 blocks from my home. Throughout this book I discussed the sacred significance of the Earth's green trees, sacred mountains, and the waters of the Earth. These 3 elements are all present around the home that I built and live in.

The Personal Feelings Of The Author and humans, the walking mummies.

The book that I wrote was certaintley not an easy task to accomplish, by any stretch of the imagination. The reseach that went into the subject of the ancient Sirius religion was extremely time consuming and physically and emotionally draining. The subject of the superior beings from Sirius, is a very highly charged and emotional issue that is the very essence of the ancient Egyptian civilization. From the ancient written artifacts that were left for modern man to examine, there is absolutely no question that man has been highly influenced by some sort of outside entity. When we examine the words of our ancestors, and compare their knowledge to the physical facts of the Earth and our world their words do in fact make a great deal of sense. When we make an honest effort to understand the many ancient artifacts that are all over the world, there is no denying the fact that our ancestors did in fact believe that they were created by the mysterious beings from Sirius. This much we can be sure of, there is no denying the fact that the Sirius religion effectively pre-dates all of the world's religions by literally thousands of years! What can be made of such a fact, in our modern world? The fact is that no one can deny that we all live on a planet that is literally covered in water, and our own bodies are also water based in their entirety. It would be silly to deny that there are not any volcanoes or granite mountains on the Earth, and that our world is not surrounded by powerful magnetic forces. It would be wrong to deny that the ancient classic style of the Greeks and Romans, was not based on the ocean and water themes of the Earth. Scientists cannot deny that Jupiter as well as the Earth are intimately involved in some sort of very mysterious electromagnetic radio contact, which at this very moment remains a complete mystery. That is why so many space probes have been dispatched to Jupiter, and why the current Galileo spacecraft is at, the giant planet. All of the themes of the Earth's oceans and water's are completely encompassing in the world in which we live, and our ancient ancestors are the ones who declare that the superior beings from Sirius are The Masters Of The Waters. Scientists today are well aware of the fact that water is quite an abundant resource in the universe. They have learned this by taking measurements with their instruments. When one looks up at the ink blackness of space at night, it does not appear that water would be a common element in the universe; however it is and it fits right into the knowledge of our ancient ancestors-that these beings are associated with the waters of the universe. The ancient Bible of the Sirius religion, which is The Egyptian Book Of The Dead, gives numerous accounts of how the Gods were born. Their Bible actually informs us that the superior God was born from nothing but invisible cold dark matter, and that he gave birth to himself! The newest scientific discoveries that are being made today do in fact agree that 95% of the universe is indeed composed of the previousley unknown invisible dark matter! In the Sirius religion we are not asked and told to have faith that God exists, but rathher we are merely informed that he does exist; and then we are told what he is made of and how he came into existence. In this religion there are no threats to the individual believer or non-believer, there is just the fact that mankind must live a good descent life on the Earth without hurting or murdering other human beings. We are informed that there is an after life, and that it is an attainable goal for every human being. For those who break the laws of the Gods however, there is un-speakable and horrible terror after death.

When we look at the fact that our own human bodies are composed of the waters and salts of the Earth, that our ancient ancestors speak of we come to the conclusion that every human being on this planet is already in a state of being mummified. Our biological bodies are naturally predisposed to being mummified, that is why a dead person who is lying in state looks like they are just sleeping. These startling results are directley related to the fact that our human flesh is salty, thus lending itself to being embalmed or mummified.

All of the facts that I have spoken of are not subject to change, consequentley our ancient ancestors knew what they were talking about. There are numerous other facts about the world in which we live, which match the ancient wisdom of man. The exact ancient and sacred numbers which are 2,3,5,7 are in fact the temperature numbers of the universe which is 2.735 above absolute zero. Here on Earth, these same numbers are also important when it comes to the very air that we breathe. For instance, the Earth's atmosphere has 23% oxygen at this very moment. If that level of oxygen were to fall down to 17%, life would soon disappear; however if the oxygen were to go up to 25% everything would become very flammable. These very numbers are in keeping with the sacred numbers of our ancient ancestors, and their relationship to the universe that we live in.

There are other very interesting facts that are associated with the sacred numbers of ancient man. The number 3 is representative of the earth, and when we closely look at this number; in the context of mankind and the various animal and sea life in the world a very interesting pattern begins to emerge. Throughout my book I have talked about the fact that mankind is intimately connected to the waters and fish of the seas, and the fact that mankind was given the dog to protect him from unseen evil spirits. Not withstanding the fact that Sirius is known as the Dog star, or

Big Dog. When we examine the 3 most intelligent and emotional species of life, on this planet, we come to the conclusion that of all the animals and the marine creatures of the Earth, only the dog, dolphin, beluga, and orca whale have emotional ties to mankind. This well known fact completely ties into the Sirius religion. Further confirming the emotional and biological connections to mankind, we know that dolphins and whales are mammals.

My personal feelings on the subject of the Sirius religion, is that our ancient ancestors were telling us the absolute truth about The Masters Of The Waters, and all that they stand for. There is no way to ignore the ancient written artifacts of man's past history, for to do so would be like trying to declare that the sun does not rise every morning. In the Bible of the ancient Sirius religion, mankind was given certain laws to live by, and those laws are not much different from the laws that other religions of the world tell their followers to live by. As I said in the beginning of my book, I am not trying to convince one single person, that the Sirius religion is authentic. Every religion of the world has positive and negative aspects to their teachings. In the world that we live, we must make extra efforts to try and understand one another; and take the time to talk with those whom we disagree with. If we cannot do this, then mankind is doomed to repeating his same old mistakes.

Through out this book I have spoken of the fish and the oceans of the Earth. The dolphin, beluga whale, and the orca whale are 3 marine animals that have emotional ties and feelings for man. Dolphins have been known to save people from sharks, and they are considered to be very intelligent creatures of the sea The same can be said of the beluga and the orca. The movie about the orca called "Free Willy" is a very emotinal account, that exposes the great attachment that exists between man and these magnificent animals of the deep abyss.

It is amazing how much emotional attachment that the marine creatures have for their human handlers. By the same token, watching these wonderful whales perform in the sea aquariums in Florida, is quite inspiring. This is the whole point of what I have been talking about in this book. The dog and the 3 marine creatures are very much representative of the superior beings from Sirius. For those who think this book is all doom and gloom, all is not lost. I base this statement on the fact that our ancient ancestors inform us that the superior beings from Sirius are amphibian, and when they arrived here on Earth eons ago they were called the Musarus which means the abomination or the terrible one to look at. Ancient man describes Oannes as terrible to look at, even scary however these beings were superior to ancient men, in their Intelligence and knowledge.

The sight of these superior beings from Sirius would certaintley be shocking to us. When they came to Earth before, they came to help mankind. Their return would also be a gesture of help to all the people of the world.

Bibliography

1.   The New Book Of Knowledge, Encyclopedia, Canada: Grolier Inc. 1966 page 482 book A

2.   Powers M. Robert, Planetary Encounters, New York: Warner Books 1978, page 249-267

3.   Sagan Carl, Parade Magazine, New York Parade Publications Inc. 6/6/93 page 4

4.   Newman Cathy. The National Geography Magazine, Washington D.C. The National Geographics Society: July 1982 page 44

5.   Websters New Lexicon Dictionary, Lexicon Publications; New York 1989 Editi on page 182

6.   Temple K.G. Robert, The Sirius Mystery, New York: Saint Martins Press 1976 page 250

7.   A Mysteries Of The Unexplained, New York: The Readers Digest: The Readers Digest Association Inc. 1982 page 49

8.   The New Book Of Knowledge, Encyclopaedia: Grolier Inc. 1966 page 378 book N

9.   Von Daniken Eric, Chariots Of The Gods, New York: G.P. Putnam Sons, 1970 page 64

10. The New Book Of Knowledge, Encyclopedia, Canada: Grolier Inc. 1966 page 214 book WXYZ

11. Von Daniken Eric Chariots Of The Gods, New York: G.P. Putnam Sons,. 1976 page 63-65

12. Langer L. William. Western Civilization, New York: Harper-American Herit age,. 1968 page 62

13. Temple K.G.Robert, The Sirius Mystery, New York: Saint Martins Press 1976 page 24

14. Kendall Timothy, The National Geographics Magazine, Washington D.C., The National Geographics Society: November 1990 page 98-124

15. Temple K.G.Robert, The Sirius Mystery, New York, Saint Martins Press,. 1976 page 11-34

16. Temple K.G.Robert, The Sirius Mystery, New York, Saint Martins Press 1976 page 250

17. Drake W, Raymond, Gods And Spacemen Throughout History, England, Neville Spearman Limited 1975 page 31 and 70

18. Kendall Timothy, The National Geographics Magazine, Washington D.C. The National Geographics Society: November 1, 1990 page 124

19. Langer L.William, Western Civilization New York, Harper-American Herita ge,. 1968 page 56

20. Budge E.A.Wallis, The Gods Of The Egyptians, New York: Dover Publications,. 1969 page 289

21. Temple K.G.Robert, The Sirius Mystery, New York: Saint Martins Press 1976 page 55-81 and page 113

22. Websters New Lexicon Dictionary, Lexicon Publications; New York, 1989 Edition page 865

23. Websters New Lexicon Dictionary, Lexicon Publications; New York, 1989 Edition page 898

24. Dictionary Of Geological Terms, A Dolphin Books; New York; 1957 page 440

25. The New Bokk Of Knowledge, Encyclopedia: Canada: Grolier Inc. 1966 page 376 book A

26. Websters New Lexicon Dictionary, New York: Lexicon Publications: 1989 Edition page 182

AUG 03

Prepare for a Martian Invasion

Backyard astronomers around the world eagerly anticipate this year's record-setting apparition of Mars. The Red Planet comes closer to Earth and appears brighter in our sky this month than at any time in the past 2,000 years. Shining at magnitude –2.9, Mars will dominate the summer sky for weeks, even outshining mighty Jupiter. Mars lies closest to Earth on August 27, 2003, when it moves to within 34.65 million miles (55.76 million kilometers) of our planet. The actual opposition of Mars occurs 32 hours later, on August 28, placing Mars at its best for centuries. What makes this opposition great is that it

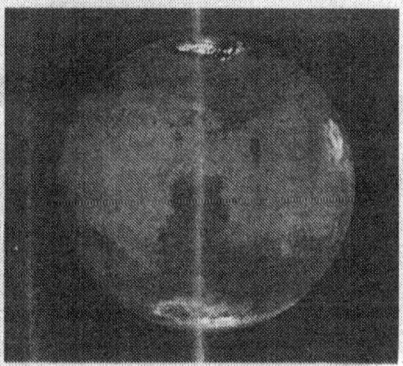

nearly coincides with Mars's perihelion, its closest point to the sun. That occurs on August 30, just two days after opposition.

Consequently, this year Mars appears the largest it has ever appeared in Earth-bound telescopes. Its apparent diameter will reach 25.13". Mars will not reach this apparent size again until the year 2208 and will not surpass it until 3282. Compare this with the previous opposition of Mars in 2001, when the planet reached an apparent diameter of 20.8" — this year surpasses that by 20 percent.

The best part of this year's martian apparition begins in May when the planet lies in Capricornus. This constellation sits too low for most observers in the Northern Hemisphere and Earth's atmosphere will adversely affect the quality of the view. The farther south you live, the better off you will be. By May 8 the disk of Mars grows more than 10". At this apparent size, Mars begins to reveal surface detail in small telescopes. Autumn begins in the northern hemisphere of Mars about May, so the northern cap will

be small. Look for the broad, delta-shaped Syrtis Major region. Also note the planet's gibbous phase. Mars stays above 10" in apparent diameter until mid-December.

The eastward track of Mars carries it from Aquarius to Pisces, and it passes south of Uranus on the way. Its eastward motion slows to a halt by the end of July.

The   Masias (Muh Sigh Uh) discovery is related to the volcanic triangle on Mars. He brought the Martian triangle to Earth and discovered the 6 global disaster triangles. 2,000 years ago when the original Messiah was in Palestine, Mars was at a close approach to Earth and now 2,000 years later with the Muh Sigh Uh discovery from Mars- Mars again approaches the Earth as it did 2,000 years ago.

About the Author

Daniel was born in Colorado and grew up in the La Junta, Pueblo and Colorado Springs areas. He attended the local school system and was enrolled at the University of Colorado. He has been involved with his Egyptian, Greek, and geologic studies most of his life. He lives outside of Colorado Springs with his wife Fawn and they have two children. Jeff is twenty four and Somer is twenty. He has built a number of homes in the Pikes Peak region and elsewhere. He did not contract them out, but as a carpenter, he literally built these homes from the foundation on up with his bare hands, without the help of other people. He continues to be involved with his global archaeological research and other activities. Right now he is working on another research book that will be ready for publication. He is the chief researcher and director of a small non-profit organization called the Global Archaeologic Research Center of Colorado. He will accept donations for this center, to continue the research, which is in service to mankind. This book is dedicated to the Gods, to wise and intelligent people around the world, and to his family.

THE NEXT BOOK COMING OUT

In my book Passage to Ascension I pointed out that our ancient ancestors believed that Gaia or mother Earth was the protector of mankind. Is this really the truth? There is astounding evidence that contradicts the ancient beliefs of mankind. In my next book I will explore this evidence and I will reveal some startling new facts about the Earth that we all live on. This research book will not disappoint you.

Daniel A Masias (Muh Sigh Uh)

The Global Archaeological Research Center of Colorado-7 Global Disaster Triangles Inc. A non profit research organization In Dedicated Service to Mankind.

For years and years Muh Sigh Uh has worked alone and with no kind of support. He has been a lonely voice. He has had very little moral support and no kind of public, corporate or government grants, or financial aid. After years of lonely research and no kind of help or encouragement from the public, his work produced an astounding global archaeological discovery. In the past 20 years he has spent thousands of dollars that he really did not have-to publish his books and print his research in the Colorado newspapers. On January 7, 1998 he published his discovery in the Gazette Telegraph newspaper in Colorado Springs. The research article spoke in detail about the discovery and the location of the 6 world wide disaster triangles. It said that in the future, disasters would take place on these triangles all over the earth. When the research came out in the newspaper, people laughed, ridiculed, cursed and spit on Muh Sigh Uh. Despite all of this criticism and persecution he continued to do his research and spend thousands of dollars to publish his new research, including this book. He held his head up high and would not bow to his tormentors because he knew that his research was true. Two years and 7 months later on July 28, 2000 he published in the same newspaper the results of the public test of his discovery, that was first published on January 7, 1998. The 2½ year public test of the discovery was 100% correct and was astounding. All around the world disasters occured in the exact physical locations where the triangles are located. Daniel performed a great service to mankind by publishing this discovery, for the benefit of man. He was the only man in the world to pay the newspaper to publish his discovery and predictions of the future, and then be 100% correct! His research center is a small room in his home as he does not have the funds to rent an office. This tiny organization is deserving of public support and donations. Muh Sigh Uh has no love of money and any book sales and donations will be going to The Global Archaeological Research Center Of Colorado- In dedicated service to mankind.

www.ingramcontent.com/pod-product-compliance
Lightning Source LLC
Chambersburg PA
CBHW030005190526
45157CB00014B/436